The Mobile Internet Handbook

For US Based RVers, Cruisers & Nomads

5th Edition

by

Chris Dunphy & Cherie Ve Ard

hosts of the

Mobile Internet Resource Center
www.MobileInternetInfo.com

Fifth Edition: January 2018

Fourth Edition: February 2016
Third Edition: February 2015
Second Edition: August 2014
First Edition: June 2013

Disclaimer: This is complicated stuff, and there are no easy, one-size-fits-all solutions. In this book, we share what we've learned over the years in our own journeys as RVers and cruisers using extensive amounts of mobile internet along the way, as well as what we've learned from wide-ranging research and conversations with other mobile internet reliant nomads.

Other than our own mobile apps, we do not sell any of the hardware or solutions mentioned in this book. We also have no financial stake in any of the companies mentioned. If you purchase anything based on the information provided here, do know that you are entering into transactions directly with the manufacturers and providers of those services, not us.

In this book we will do our best to share the pros and cons of each option as we understand them today, but ultimately you must continue your research and decide for yourself what solutions fit you best.

We can take no responsibility for the choices you make as a result of reading this book.

When issues arise, please seek technical support and resolution from the provider, vendor, or manufacturer you purchased from – not us!

The subject of this book is constantly evolving.

It's ~~almost~~ guaranteed that as soon as we publish this book, there will be an industry development that makes something in this book out-dated.

That's why we also host MobileInternetInfo.com, the website meant to complement this book. We are constantly staying on top of this topic - analyzing tech news for nomads, updating in-depth guides, reviewing products, creating videos and more.

Mobile Internet
RESOURCE CENTER
mobileinternetinfo.com

For free periodic wrap-up articles that update this book:
www.MobileInternetInfo.com/changes

Other places you can keep updated:

- Join our free monthly newsletters:
  www.MobileInternetInfo.com/subscribe

- Join our free public Facebook discussion group:
  www.facebook.com/groups/rvinternet

- Follow our free Facebook News Page:
  www.facebook.com/MobileInternetInfo

- Add our news feed to an RSS reader:
  www.mobileinternetinfo.com/feed/

- Subscribe to our YouTube Channel:
  www.youtube.com/MobileInternetResourceCenter

Dedicated to

Tim VeArd
1944 – 2013

*Technology pioneer, national hero, and
location-independent technology
entrepreneur who inspired us in ways beyond
imagination.*

We miss you, Dad.

*And a HUGE Thank You to all of our
supporters who pre-purchased the 2014
book edition, funding our first major re-write,
and who made it possible for us to launch
MobileInternetInfo.com - turning this topic
into our our full-time focus!*

Table of Contents

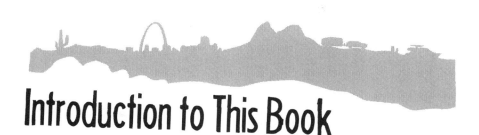

Introduction to This Book

For anyone thinking of exploring a mobile lifestyle, figuring out how to best keep **online** while exploring the world **offline** becomes critically important.

Whether you use the internet to keep in touch with loved ones, plan your next destination, manage your finances, learn online, pursue entertainment, or working remotely – building a connectivity arsenal that suits you is an essential chore most tech-connected nomads face.

Even those who don't consider themselves techno-savvy at all still face needing at least some connectivity these days.

Prologue to the 5th Edition

After years of sharing about our own RV travels while working remotely, we set out in the spring of 2013 to write a comprehensive blog post about the best options for keeping online while mobile.

Our prime goal was to slow the flood of questions we kept fielding.

By the time we were partway through, we realized this topic was way too big to be covered in a single post!

Introduction to This Book

And thus, on the spur of the moment, the first edition of *The Mobile Internet Handbook* was born. We had never published a book before, and we put together the first edition in under three weeks on a shoestring budget, eager to get back to our *real jobs*.

We did not expect the book to take on a life of its own -there was clearly a huge need for unbiased mobile internet advice.

In mid-2014, we updated and vastly expanded the book and we also launched a companion member-supported website which has now become a comprehensive online resource center, originally focused on RVers.

We have been thrilled that enough folks have proven willing to join a premium membership tier to support the ongoing work through now three additional book editions, and the creation of our ever-expanding library of in-depth written and video content.

**We are proud to be reader &
member funded - thank YOU!**

Now this brings us to this new fifth edition, being released in 2018.

The mobile industry continues to evolve at a rapid pace, and though we had written the 2016 edition of the book to be as evergreen as possible, after nearly two years an update was essential.

This new edition takes the book back to its roots, more of a handbook-sized overview than the comprehensive, overwhelming and quickly outdated standalone resource that prior editions of the book had become.

This edition focuses on explaining the basics, helping you assess your needs, and laying the foundation for your continuing research. To go further, we've provided links to accompanying guides on the MobileInternetInfo.com website that we keep current with the latest options.

Also new in 2018 - we've expanded our target audience beyond just RVers. We are now focused on all US-based nomads traveling by land or water!

Who This Book Is For

This book focuses on connectivity options for RVers, cruisers, and nomads - whether full-time, seasonal or those on extended excursions.

If you are setting out to explore the vast expanse of the USA for a prolonged period of time and want to remain connected while doing so, we've written this book for you.

Introduction to This Book

We aim to present information in a clear manner that will be accessible to most, ranging from novices who don't yet understand the difference between Wi-Fi and cellular, to the fellow super geeks.

This book isn't intended to be consumed in one sitting, nor do we expect that you'll understand it all immediately. Dive in at your own pace.

There is no denying that at the root of it all, *The Mobile Internet Handbook* covers pretty technical stuff. We try to start each topic off as fundamentally as we can, and then ramp up from the basics.

Armed with the information in this book, you will be much better equipped to understand and evaluate the current offerings on the market – and to decide what plans and technologies personally fit you best.

It is not our goal to give you a shopping list for the singular perfect mobile internet setup. Our goal here is to arm you with the information you need to research and create your own.

Keep in mind, you're going to be out there needing to manage whatever connectivity toolkit you assemble, often miles away from any geek help.

And, if you are having trouble getting online, you may end up being unable to ask for assistance in online forums and groups.

So, before you get lost in the wilds (figuratively or literally), make sure you understand what you have and how to use it!

This handbook is here to help lay that foundation.

The Mobile Internet Resource Center and Membership

The world of mobile internet is constantly evolving. There are always new plans, products, and options coming onto the market.

With as often as things change, it's impossible to keep a comprehensive book current. Even annual updates would not be enough.

So instead of cataloging specific

Mobile Internet
RESOURCE CENTER
mobileinternetinfo.com

plans and products that will immediately be out of date, this book is just a primer.

The companion website MobileInternetInfo.com is focused on the things that are constantly changing, sometimes on a daily basis. There we track the current plans and products.

We have news articles, dozens of very in-depth guides, video content, product overviews, hands on tutorials, field testing data and much more.

We've now also launched a new resource for our members - the Mobile Internet University. This online classroom goes along with this book, walking you through our content to help assemble your own unique mobile solution.

The book and online content are intended to complement, not replicate, each other.

Out team works full time to track the industry and keep our guides updated, and our deeper in-depth guides & product reviews have hundreds of hours of research, testing and content creation invested in them.

Sales of this book alone don't begin to cover that massive effort.

So instead of seeking corporate sponsorships and advertising, we have chosen to be community funded via premium memberships. This way, we are accountable only to you, our readers and our members.

In return for their support, premium members get this book and even more resources, such as interactive access to us via our forums and webinars, expanded video content, even more in-depth guides, hands-on product reviews, access to our online classroom, and special alert newsletters.

Some of the guides you'll find linked to from this book might contain a mix of free and member-exclusive content, so do be aware that your purchase of this book does not give you access to everything we offer online.

We strive to draw the line between free and member content so that most casual mobile internet consumers will find everything they need without needing to join.

This book is written to be informative on its own, and is not meant to be a hard sell pushing the membership.

But for those for whom mobile internet is essential, there is a lot to be found by diving in deeper with us. Come join us when you are ready and please accept the discount code we've offered at end of the book In thanks for your purchase.

Your Authors: Chris & Cherie

We're Chris and Cherie, also known as the Technomads of Technomadia.com.

We gave up our fixed based homes and began traveling full time in 2006.

Currently we split our time between RVing the US in our vintage bus conversion, and cruising America's waterways in our boat as we slowly explore America's Great Loop.

We hit the road when we were in our early 30s, too young to be traditionally retired and not fortunate enough to be retired early.

And we have some serious wanderlust.

Thankfully, mobile technology allowed us to take our high-tech careers in software development, strategy advising, and technology consulting on the road, working remotely every step of the way.

We formed our business Two Steps Beyond LLC (www.twostepsbeyond.com) to do all of our work projects under - including now hosting MobileInternetInfo.com.

We consider ourselves to be "technomads" and have been able to create a lifestyle that combines our careers and our desire for mobility.

We know this topic intimately because we absolutely depend on mobile internet to keep connected to our clients, manage our projects, keep in touch with loved ones, and make new friends on the road.

And well, we're just geeks who have been surfing online since the mid-80s. We just can't imagine living life offline for long.

About Chris: Before going nomadic, Chris had a career in Silicon Valley focused on mobile technology.

He started out as the founding Technical Editor of *boot Magazine*, mastering the art of explaining complex topics to a mainstream audience.

Introduction to This Book

His most recent corporate job before going nomadic was as Director of Competitive Analysis (aka "Chief Spy") for Palm and PalmSource, the companies behind the pioneering Palm Pilot. His job was to be intimately familiar with every mobile device and technology in existence, and he was tasked with traveling the world to dig up information to chart the future of the mobile industry.

It was not uncommon for him to be carrying dozens of mobile devices with him at a time, always raising eyebrows passing through airport security scans.

Times haven't changed much actually - we still have dozens of mobile devices on board!

About Cherie: Cherie began running a software development business from home in the mid-1990s, and has a long history of working remotely for her clients, including pursuing personal travel adventures.

Aside from developing custom business software, her career involved technical writing and teaching high-tech topics to non-technical people.

When she met Chris during his first year on the road, she was well accustomed to already carrying smartphones able to tether her laptop to the internet.

Devising a reliable mobile internet solution was essential for her to join him full-time on the road while still running her business.

Together, our first year on the road was spent in a tiny teardrop travel trailer equipped with just the essentials – solar electricity and mobile internet. We've switched it up several times since then in various modes of travel.

Introduction to This Book

Covering mobile internet is a perfect meshing of our passions for technology, mobility and helping people pursue their dreams.

It has now become our professional focus.

It is truly an honor to provide this resource to you.

Follow our personal nomadic adventures:

Blog:
www.technomadia.com

Facebook:
www.facebook.com/
technomadia

YouTube:
www.youtube.com/Technomadia

Instagram:
@cherie_technomadia / @chris_technomadia

And, our personal mobile internet setup:
www.technomadia.com/internet

Laying the Groundwork

Let's get right to the question we know you bought this book to answer:

> What's the best way to get online while traveling?

The answer is...

> ... whatever works best at your current location.

And that is the challenge of a mobile lifestyle. Your location is always changing, so the best way online may be changing too.

Some days a cellular hotspot with a direct wired antenna might be best.

But at your next stop you might find workable guest Wi-Fi offered, or be able to subscribe to your own cable internet hook-up.

And in some places you might need to switch to a different cellular carrier, or use a booster, or even a directional antenna - all just to get even the slightest hint of signal.

And sometimes, relying on satellite internet may very literally be the only option for miles around.

Or you might have to relent and commute into town to find Wi-Fi at a local library or cafe.

In other words, there is no one single "best way."

Laying the Groundwork

The key is striking the right balance between your needs, budget and sanity.

To do that, you need to understand the challenges of mobile internet, your specific needs and the trade-offs of the options available.

Basic Differences Between the Common Options

There are multiple ways to access the internet while on the go, and each of them has attributes that might make one more attractive than the others.

Here is a quick grid comparing the primary options:

	Unlimited or Capped	Mobile Friendly	Cost	Speed	Reliability
Cable/DSL	(Usually) **unlimited**	Not mobile - fixed location	Reasonable	Fast	Always on
Cellular	Often "unlimited" with caps & management	**Fully mobile – wherever there's signal**	Reasonable to pricey	Slow to faster than cable	Variable
Public Wi-Fi	Variable	You hunt signal at each location	**Free to cheap**	Highly variable	Highly variable
Satellite	Capped	**Mobility can be limited, but if allowed - works in even the most remote places**	Pricey to beyond pricey	Varies, with latency issues	Variable

- **Cable/DSL:** This might be what you're used to at a stationary home, but on the go, this option is only sometimes available in places that cater to longer-term stays.

- **Cellular:** Cellular data is quite prevalent and has gotten amazingly fast. It is now even available in sometimes surprisingly remote locations. While many carriers now offer unlimited data, you'll quickly find there are lots of limits to "unlimited" to understand. You might also need extra equipment to optimize utilizing cellular in remote locations. You'll need to select your carrier(s) and equipment wisely to best match your planned travel destinations and routes.

- **Wi-Fi:** Public Wi-Fi hotspots are often free or low cost, but they can vary vastly in quality and are frequently too overloaded to be reliable. Unless you're willing to take your laptop physically closer to the hotspot, you may need additional gear to get a usable signal from the comfort of your RV or boat.

- **Satellite:** Satellite service can be picked up anywhere with a clear view to the southern sky, even in the most remote locations. But, satellite internet options typically come with a host of drawbacks.

More than likely, most nomads will create a personal arsenal that combines multiple options to best fit their own unique needs.

We'll overview each of these options later in the book.

What Exactly Is Data?

Just like liquids are measured by the drop, teaspoon, cup or gallon - data is measured by how much storage space it takes up.

Digital data is made up of bits – literally, zeros and ones (0 or 1). Think of each bit as basic molecules of water.

It takes eight bits to make a byte. And a byte is generally a single character of text. For example, the letter 'A' is stored and transmitted as 01000001 in binary code. For simplicity, let's say a single byte/letter is like a drop of water in our liquid analogy.

It takes 1024 bytes to make a kilobyte (KB), which is roughly 1024 letters - or a decently sized paragraph of text. Continuing our liquid example, 1024 drops of water equals roughly about 10 teaspoons.

Combine 1024 kilobytes and you get a megabyte (MB). Just like if you combine 1024 teaspoons you'll get about 20 cups of fluid.

Add together 1024 megabytes and you make a gigabyte (GB or 'gig'). 1024 cups of water would equal about 64 gallons - the size of decent fresh water holding tank on a RV or boat.

Cellular and satellite internet service is often sold by the gigabytes per month. We're not talking about how much data is necessarily stored, but how much is moved over the internet from point A (for example - Netflix's server) to point B (your smartphone).

Even "unlimited" plans can have high-speed usage caps that you need to be aware of. Your usage may be technically unlimited, but after consuming a certain number of gigabytes your speeds might be slowed drastically for the remainder of the month.

How much stuff does a single gigabyte equate to?

Laying the Groundwork

Text takes very little space. A book like this one (stripping away images) is only a few hundred thousand characters, less than 400KB - or using our analogy, about 8 cups of water. Data compression techniques work incredibly well on textual data, making text take up even less space in practice.

Because text is so efficient, email (without attachments), messaging, and even basic text-centric web surfing uses so little data that it is hardly worth worrying about unless you are on a VERY limited connection. This sort of digital activity is sort of like normal daily drinking from a fresh water tank - you really don't need to stress about running out.

Pictures and music, however, start to move the needle a bit.

A typical consumer-grade digital camera or smartphone takes photos that average around 2.5MB in size. In other words, when it comes to data, pictures are actually worth *several* thousand words, especially if you don't reduce their size before emailing or posting them!

It takes around 400 images of this size to equal one GB. The modern web has grown very graphically rich, and web pages can easily consume 1MB to 5MB per page viewed, and sometimes more.

This means a GB of mobile data may only allow you to load 200 web pages.

Streaming music or podcasts online can consume 30MB to 90MB or more *per hour*, depending on the audio quality. A GB of data can equal about 11-35 hours of audio streaming.

Video is the ultimate data hog - the equivalent of taking long showers and doing your laundry in terms of water usage. No matter how big your tanks, you need to be careful or you will run out!

Especially with high-definition (HD) video, data amounts quickly begin to be measured in *gigs per hour*.

A 90-minute movie streamed in HD can easily consume 4-5 GB of data, and a 4K ultra-HD stream can burn through more than twice that (the equivalent of filling a hot tub in our water usage example)!

Reduce the resolution down to standard 480p (DVD quality) or even lower within the settings of your streaming service, and video is now less than a GB an hour. Think of this like using a water saving shower head and turning the water off while you soap up.

But stay on guard! If you are on any sort of limited data plan, even low-res video can drain your data tanks quickly.

Hidden Data Hogs

There are plenty of things you can do on a computer or smartphone that will not count against your monthly data usage.

But, there are plenty of other things that consistently catch people by surprise.

If you are on any sort of limited connection, it is important to understand just what sort of hidden hogs you might have lurking.

For example, if you use a cloud syncing service (such as iCloud Photo Library or Google Photos) to share your photos between devices, that can use substantial amounts of data copying ALL your photos to the cloud and back - potentially multiple times.

And, if you sync backups online or use a service like Dropbox, every time you make even a small edit to a file you may be inadvertently triggering substantial unintended usage as the full updated file gets backed up and then copied down to your other devices.

Be Aware of Activity Behind the Scenes

Regardless of what you are explicitly doing in the foreground, modern computers, smartphones, tablets, and other devices are often busy behind the scenes burning data that you might not be aware of.

So many devices now run automatic updates or rely on cloud syncing of data, and it seems designers universally assume everyone has access to a fast unlimited home internet connection.

For more on tracking down Data Hogs:
www.MobileInternetInfo.com/datahogs

Rule #1: Reset Your Expectations

The technology for connecting while on the go has advanced at an incredible pace over the past few years. There have been vast improvements in both data speeds and coverage, and it's only getting better.

Meanwhile, the price for mobile internet data has plummeted, though typical monthly usage has gone up even faster than prices have gone down, so things might not feel any cheaper.

Sometimes, you can even get connection speeds while on the go that exceed what you could get via the fastest fixed-place connections.

It is downright amazing at times.

But, despite all these advancements, there are still limitations and plenty of frustrations.

The most important thing you can do to prepare yourself is to reset your expectations.

> *Be ready for the bad days.*
>
> *The slow days.*
>
> *The no days.*

We're not trying to scare you away, but we do want to make sure your expectations are realistic.

Keeping online most of the time while traveling is entirely possible, but it's not necessarily always easy, cheap, or with the snappy fast speeds you might be used to.

Navigating mobile internet is not going to be anywhere near as easy as just plugging in a cable like one you might have had in your fixed home. And no, they don't make a cable long enough to take your wired internet with you.

You will be battling:

- Intermittent and unreliable connections.

- Varying speeds - from frustratingly slow to blazingly fast.

- Data usage limitations and caps.

If your mobile livelihood absolutely depends on keeping connected, you will have to carefully plan your mobile life around this need.

This could mean altering routes and carefully planning where to stop to get work done.

It may mean having to move on sooner than you're ready to search for better bandwidth, and may even mean afternoons spent at the local Starbucks, McDonald's, brewery, or library to soak up some Wi-Fi.

Laying the Groundwork

If you can turn the inevitable frustration around from "Gah! I have to go drive to find Wi-Fi" to "Oh darn, I have to indulge in a local craft beer while I get some work done," you'll be better able to thrive in this lifestyle.

No matter how many backup plans you build into your connectivity arsenal, when you absolutely need to get online, that's inevitably when the glitches will emerge.

No matter how much time and money you invest in ensuring great connectivity, there will be times that you can't get a stable connection to do everything you need from the location you want to be at.

Mobile equipment can (and will) fail, firmware patches and upgrades can cause unintended problems, weather can interfere, or your exact location can influence your signal. Even something as invisible as your neighbor's microwave oven can conflict with your Wi-Fi network, knocking you offline until their popcorn is ready.

You need to realistically set your expectations, as well as the expectations of the people depending on you being online, such as clients, co-workers, family, and friends.

> There will inevitably be compromises in connectivity in exchange for your mobility.
>
> **What tradeoffs are you willing to make?**

Very seriously consider the costs of assembling your mobile internet options. Decide for yourself how much it is worth spending to try to cover your bases, and how much internet access *you really need*.

Even if your income source does not depend on internet access, you may need to adjust your expectations around how much you rely on connectivity for personal reasons such as email, Facebook, viewing video content, banking, bill paying, looking up information about your next destination, online learning, gaming, and keeping in touch with loved ones.

If you're used to streaming TV and movies over your fast unlimited cable internet, you may need to adjust your viewing habits to include other sources of entertainment.

Laying the Groundwork

If online gaming is your entertainment of choice, you may need to resign from your Warcraft clan and focus on single-player or turn-based games instead.

And, if you've gotten hooked on video chatting, you may need to actually resort to old-school voice phone calls every so often as a backup.

No matter what you do there will be days that staying connected is more of a headache than it is worth. The most important rule for staying connected on the road is that you need to be prepared for these days.

Companion Online Glossary

If at any point you come across terms in this book that are at all unfamiliar to you, please check our free online glossary.

Our Online Glossary:
www.MobileInternetInfo.com/glossary

This online glossary is intended to define even the most technical terms in ways that most should be able to understand.

So, before you get frustrated wondering why you might need a POE to power your CPE to get remote 802.11g when you'd really rather have more dB on your LTE – check the glossary, and soon it will make better sense.

Assessing Your Needs

There is no one single technology for keeping online that is appropriate for all the different situations mobile users might find themselves in.

Rarely do we see two mobile households select identical setups.

Your ideal arsenal is going to be very personalized to you and dependent upon several factors related to your travel style, internet needs, budget, and technical comfort level.

In creating your arsenal, consider the following questions...

Personal Considerations

How important is internet access to you?

Do you NEED to be online during certain hours of the day to work or attend classes? Are you addicted to social media, streaming video, or playing online games?

Or will your mobile internet access needs be more flexible?

If you won't get the shakes (or lose your job) if you can't get online today - or heck, even this week - then you might not need as many options onboard as someone who absolutely must get online by a certain time every morning.

If your needs are more in the 'absolutely essential' camp, then you'll want to plan your setup to include multiple levels of redundancy and signal enhancing solutions to ensure better odds of getting online.

What do you need to do online?

The mobile internet setup of someone who needs to handle high-bandwidth streaming, two-way video conferencing, or manage large files remotely will look very different than someone who just needs to check e-mail, plan the route ahead, and manage their finances.

And of course, the data needs of an entire connected household with multiple computers, tablets, streaming devices, security cameras, and gaming machines to keep connected is vastly different than that of solo nomad just doing some casual surfing on a single tablet.

If your data needs will be high, then you'll want to seek out unlimited data options, signal enhancing gear, and perhaps even a dedicated router to better share the connection.

For the more casual surfer, you may get by just fine with a small data plan on your phone and soaking up free Wi-Fi when you find it.

Whatever you do, avoid putting together a needlessly expensive and complex system. Overkill is rarely the right answer.

Check the **Use Cases** chapter at the end of this book for some considerations of common things folks need to do online.

What is your style of travel?

Oh, the places you can go. That's the whole point of a mobile lifestyle!

If you're planning to hop between urban RV parks or marinas, you'll probably be staying in areas with great connectivity. You'll likely be able to utilize cellular data plans from any of the major carriers and you will likely even have access to public Wi-Fi hotspots.

If you prefer going more off the beaten path, such as to amazing state parks, national parks, or heading off into the wilds, then connectivity will be much more challenging.

To stay connected in more remote locations, you'll need to put together a solid signal enhancing strategy with multiple options for getting online. Usable public Wi-Fi will probably be scarce.

If you're really going off into the boonies, you may even need to consider satellite internet.

Another consideration is how often you plan to move locations.

Assessing Your Needs

If you're staying places with extended stays measured in months or even years, then you can optimize your setup for what works best in that exact location.

But if you'll be hyper-mobile and changing locations frequently, you'll want to consider multiple carrier options and devices to handle diverse situations. And also, optimize around easy-to-deploy gear.

What plans and devices do you have now?

Take into consideration what your current mobile internet setup is. Some folks still only have flip phones (or *gasp* - land lines) and are truly starting from scratch. Most of us these days already have a smartphone or tablet.

Is your equipment due for replacement soon? Or do you want to maximize your current equipment investment?

Do you have contracts on your cellular plan and devices that would be costly to break?

Keep in mind, technology changes rapidly and it is likely that you will want to re-evaluate your setup every year or two to stay current.

Think it through carefully before getting locked in.

How comfortable are you with technology?

There's no denying it - mobile internet is high tech stuff. It is easy to get overwhelmed and left feeling like you need a full time geek living onboard.

If you're not comfortable with technology, keep it simple and stick to stuff you understand and can manage on your own. You can always add more options later as you increase your comfort level.

For those who are more technically inclined, consider how much time you want to invest in being an IT manager.

Even some of the most technically advanced users are happy to pay for simplicity, reliability and even hand-holding support just so they can have someone to call when things don't work right.

Other advanced users thrive on complexity, and enjoy geeking out and building a custom one-off personalized setup.

What's your budget?

The cost of staying connected can add up quickly between upfront equipment purchases and monthly fees for plans.

Don't add every product category to your set-up until you truly understand if it's a fit.

Throwing money at this stuff doesn't necessarily keep you connected.

When are you Hitting the Road?

If you're not hitting the road in the next couple of months, please don't jump into buying all your equipment right away!

Technology changes so quickly that you are best off leaving the final assembly of your connectivity arsenal until much closer to when you are actually hitting the road. We recommend no sooner than 6 months before heading out, but 1-3 months before is even better.

How Much Data Do You Need?

Mobile data, while sometimes marketed as unlimited, tends to have restrictions on usage. Getting a handle on your actual expected usage is critical when deciding what you need.

To start with, you should do an assessment of your usage and monitor it for a while - even before you hit the road.

If your internet provider doesn't provide a monthly usage number for you, it is recommended that you install a counter on your computers and/or router to record your monthly usage. Most mobile devices have built in tracking available in the settings menu.

If you can't find a tracking solution that works for you, some folks opt to purchase a no-contract cellular data plan to get a handle on their data needs.

Regardless of your method, track your actual regular usage for a reasonable amount of time (at least a full week or, better yet, a month) to get a baseline. Remember to factor in all of the devices that you'll be connecting on the road - tablets, smartphones, music players, game systems, eReaders, laptops, streaming devices and computers that will be in your household.

Also consider how you currently consume media content like movies and TV shows. If you're doing that over cable TV now, how will that translate for you once you no longer have cable?

Now, compare your own personal baseline to what it would cost to buy that much mobile data.

You may be shocked by these numbers, and have to weigh the balance of the cost of mobile internet versus trimming back your usage.

Assembling & Organizing an Arsenal

Unless your needs are very basic or you'll be fairly stationary in one spot, you'll likely want to consider multiple solutions for keeping online to try at each stop you make.

When Plan A is out of range or overloaded, Plan B suffers a hardware failure, a tree is blocking the signal to Plan C, and you ran over the wire to Plan D – what will you try next? How much redundancy do you need?

Sometimes what works best in a single location will change based on the time of day, or the weather.

But getting this much gear and plans together requires some planning and time to bring together.

Assemble Over Time & Re-Evaluate

Don't feel you have to have your total connectivity solution built on day one.

Yes, some things will be easier to install when you have access to ladders and tools and trusted installers. But, you can always change things up later if you find something isn't working for you, or you decide to change up your style of travel.

And as much as we'd love to tell you that you only have to go through this process once, technology will continue to advance. New plans will come out and new gear will be launched.

If internet will play an important role in your mobility, keeping current will help keep you better connected.

Be prepared to re-evaluate your setup every year or two. Keep your setup as flexible as possible by considering the costs of the equipment you purchase and any service contracts you might enter.

The Tech Cabinet Approach

Now, consider what to do with all this gear you might be bringing together.

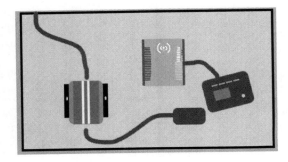

A tech cabinet is a neat little place where you keep all of your tech together - kind of like a server room/closet or a computer room/office area in a house or office.

And it's a great strategy for bringing all your gear together in one place to optimize your signal enhancing, wiring, power and ventilation.

Since mobile dwellings (i.e. RVs, vans, boats) are typically smaller than a house or office, most folks don't have the space to dedicate an entire room to storing tech gadgets.

Instead of dedicating a room to the task - it is often ideal to set aside a cabinet for this task.

For more on the Tech Cabinet Approach:
www.MobileInternetInfo.com/techcabinet

And never forget that trade-offs will be inevitable. It may be frustrating in the moment, but the occasional bout of disconnectedness is a small price to pay for all the incredible perks of a mobile lifestyle!

For more on Assessing Your Needs:
www.MobileInternetInfo.com/assessing

Going Wireless

As you shift from stationary internet to mobile internet, you're going to need to come to grips with the realities and unique challenges of wireless signals.

Wired connections like you might be used to with cable or DSL tend to either work well, or not at all.

Wireless signals, on the other hand, can be fickle. To put it mildly.

That solitary bar of signal is a cruel tease, usually not enough to actually reliably use, but it's there - taunting you.

If you maybe just hit reload one more time... or shift your position a bit...

Right at the moment you are about to give up, it works! For a few minutes at least. Just long enough to keep you on the hook trying.

At least when there is no signal at all, you can concentrate on doing other offline things. But having a hint of signal... that is the path to madness and investing hours accomplishing nothing.

There are, however, things you can do to improve a bad situation - taking a weak signal and making the most with it. With external antennas, cellular boosters, and long range Wi-Fi gear, the results can sometimes be near miraculous.

Other times, no amount of enhancing can help. That cruel solitary bar might remain unusable, taunting you still.

With some basic understanding of how wireless signals work, you can optimize things on your end to ensure you get the best possible performance.

Understanding the Difference Between Cellular & Wi-Fi

The two most common ways that mobile users utilize to get online are either a cellular data connection or a public Wi-Fi network.

Both are wireless signals, so just how are they different? And why do they need separate antennas and booster equipment?

Here is the basic breakdown:

Wi-Fi

This is a short-range local wireless network technology used for connecting Wi-Fi enabled devices to each other. It's not necessarily the internet itself, but is often used in this manner in casual conversation. You might say "I need to find some Wi-Fi" when you really mean "I need to get online."

A Wi-Fi router, when connected to an internet upstream connection (such as cable, DSL, satellite, or even cellular), can also share an internet connection to its connected devices. This is all done via a wireless signal that can generally be received in a limited area - typically only few hundred feet away.

All modern laptops, smartphones, tablets and most other internet connected devices have Wi-Fi receiving ability.

Many Wi-Fi connections you encounter will be provided for free - offered by campgrounds, marinas, cafes, stores, libraries and hotels as a perk for their customers. But there are some paid options out there like Boingo, TengoNet and Xfinity too.

To further add confusion, a Wi-Fi hotspot may also be one you create and host yourself, such as a hotspot off of your smartphone, mobile hotspot or router. In those cases, you may be paying for the internet data being utilized.

Cellular Data

Cellular is a longer range wireless technology. It is the same basic technology used for cellular phone calls and text messaging.

Each cell phone company (aka carrier) builds cell towers in locations it has customers to serve, and each tower transmits a signal of varying power that can be picked up by devices within its coverage zone - called a cell.

Those individual cells may range from the size of a city block to the size of an entire town (or larger!)

All of these cells are coordinated carefully and designed to overlap, creating a network of coverage.

As you move out of one cell, the connection to your device is handed off to the tower serving the next cell, usually fairly seamlessly. The places where there are gaps between the cells where no tower reaches are known as dead zones.

The quality of your connection will be affected by how far from a cell tower you are, obstructions that may exist between you and a tower, and how many other customers are utilizing the same tower you are connected to.

All smartphones, some tablets, some newer cars, a very few laptops, and some routers have cellular data capabilities built in.

When using cellular data you are accessing the internet via a cell tower that might be within sight or perhaps as far as 20 miles away. If you have a weak cellular signal, sometimes cellular antennas and cellular boosters can help you improve the signal.

Cellular data is rarely free, and access requires you have a data plan with a cellular carrier such as Verizon, AT&T, T-Mobile, or Sprint.

The radios involved, frequency bands used and technologies underlying Wi-Fi and cellular data are different. A cellular booster cannot help you pull in a Wi-Fi signal from farther away. Extended-range Wi-Fi equipment will do nothing to improve your cellular reception.

In addition, cellular and Wi-Fi radios are completely different from and not compatible with TV antennas either.

Things That Impact Wireless Signals

Wireless signals can be quite variable and are influenced by many things. Sometimes just moving from one side of your couch to the other can make a dramatic difference in the reception of a signal.

In general, there are a few universal strategies that will help with reception, relevant for receiving both cellular and Wi-Fi signals.

Things that help:

- **Line of Sight:** Nothing improves a wireless signal more than having nothing between the sender and receiver. If you can visually see the cell tower or the Wi-Fi hotspot, there is a good chance your wireless device can too.

- **Altitude:** The best way to get a line of sight, or at least fewer obstructions, is to get up over the clutter. An antenna at the top of a 20' mast has nothing but clear air around it. An antenna mounted directly on the roof might be blocked when a giant fifth-wheel moves in next door, or even by your own air conditioner. An antenna on any device stuffed into the back of a crowded lower cabinet is starting off with a substantial handicap.

- **Directional Focus:** A transmitter of a given loudness (power) can be heard a lot farther away if the energy is focused towards the receiver, just as speaking into a megaphone does wonders to project your voice. The downside is a directional antenna requires manual aiming. Since the signal off to the sides outside of the sweet spot is substantially diminished, directional antennas might be more trouble than they are worth if you're just passing through. In a relatively fixed location the extra aiming effort can be worthwhile. Boats swinging at anchor make directional antennas completely impractical.

- **Power:** Louder is better - but only up to a point. If you push too much power out of a radio antenna, it can begin to overwhelm your target. And worse, it can drown out others using the same or nearby frequencies. Maximum power output is usually capped by FCC regulations.

- **A Sensitive Listener:** Just like in relationships, a sensitive ear is often better than raw power. If the receiver is listening carefully, even a weak signal might be heard. This sensitivity is often a prime difference between expensive commercial-grade wireless gear and cheap consumer-grade stuff.

- **An Uncrowded Channel:** Wireless networking devices are designed to allow multiple users to share a single channel. On a cell tower or a crowded Wi-Fi network, there could be hundreds of devices all trying to talk at once. The more people talking at once, the more congested and degraded the network becomes for everyone.

- **Upstream Capacity:** A river can't flow through a straw, and often the real problem with both public Wi-Fi hotspots and cell towers is that there's only a straw's worth of pipe bringing in the water. Some campgrounds and marinas invest in great Wi-Fi equipment, and you might have the strongest signal ever. But, upstream they may have little more than a single basic DSL line serving the entire facility, rendering the actual Wi-Fi experience frustrating. The same concept applies to cell towers, particularly in more remote areas. The tower may speak fast LTE, but if the upstream network is not sufficient for the population being serviced, the actual speed you see may be extremely limited despite a seemingly great signal.

- **Antenna Diversity:** Having multiple antennas working together is called antenna diversity. This can work wonders with compatible equipment, particularly in areas where there may be a lot of signal reflections bouncing around. A secondary antenna can compensate for a primary antenna being in signal shadows, and also enables MIMO technology (see below) to kick speeds into overdrive.

There are some universal things that degrade all signals. Things that hurt:

- **Metal Obstructions:** Barriers between you and the Wi-Fi base station or cell tower are bad. Metal barriers are very, **very** bad. Radio waves in general have difficulty passing through metal and are often reflected away. This makes getting a signal for those living inside metal buses, shiny Airstreams, or steel-hulled boats an extreme challenge.

- **Interference:** The more noise the harder it is for a radio receiver to hear. Noise can originate with other intentional signals or it could be background noise from microwave ovens, hair dryers, or even the sun.

- **Overcrowding:** Overcrowding leads to a vicious spiral. With too many radios trying to broadcast on one channel it becomes hard to pick out individual conversations. Transmitters try to compensate by broadcasting louder and repeating themselves to get a message through. That leads to even more overcrowding and noise until eventually the wireless network saturates and data stops flowing.

- **Power**: Some Wi-Fi devices and routers let you manually set your radio power. It seems counterintuitive, but you can actually sometimes improve both speed and even range by reducing the power setting. At full power, you may be too loud, and you could be generating interference and overcrowding your neighbors. Speaking in a whisper is sometimes more effective than roaring at the top of your lungs.

A lot of these things may be out of your control, but it helps to understand the things that may be impacting your signal in any given situation.

What on Earth is MIMO?

MIMO (Multiple Input, Multiple Output) is one of the core technologies enabling both Wi-Fi and 4G/LTE cellular. Almost every cellular LTE device has at least TWO cellular antennas to enable the magic of MIMO.

Think of it like stereo hearing - using two ears to listen with instead of one.

In weak signal areas, MIMO lets the radio combine the signal to better pick up the data stream from the background static.

And in stronger signal areas, MIMO actually enables a special turbo-mode, doubling speeds by allowing each antenna to have its own connection with the cell tower.

Some of the latest cellular devices even have FOUR antennas now, allowing for the possibility of quadruple-speed MIMO!

MIMO is not just for cellular. Many of the latest high-end routers, laptops, and desktops support up to 3-way Wi-Fi MIMO, using three transmit and three receive antennas working together to give peak Wi-Fi data speeds. Some manufacturers may not refer to this as MIMO, but instead just refer to how many antennas are built in, or they may use terms like 'triple chain'.

However it is described, the more antennas the better.

For our Guide to Understanding MIMO
www.MobileInternetInfo.com/mimo

Knowing Where to Find Signal

A critical part of successfully navigating a mobile lifestyle is knowing where along your routes you'll have the best chance of getting a signal.

For those who rely on mobile internet, it can be very important to have this information before you head out to a new location.

Thankfully, there are tools available to help, whether your goal is to connect via cellular or Wi-Fi.

It pays to do some research in advance!

Checking the Cellular Carriers' Online Maps

All four of the major cellular carriers publish their coverage maps online.

If you have service via a reseller, they may not have a coverage map online. But, if you know who the underlying network provider is, you can go right to the source.

- **Verizon** – www.verizonwireless.com/wcms/consumer/4g-lte.html

- **AT&T** – www.att.com/maps/wireless-coverage.html

- **T-mobile** – www.t-mobile.com/coverage.html

- **Sprint** – coverage.sprint.com

Coverage? App

Although we can go to each carrier's maps online separately, we decided to make it easier by bringing the four major carriers' maps to your smartphone or tablet. Yep, we wrote an app for that!

Coverage? (available for iOS and Android) allows you to overlay regional resolution versions of the carriers' maps so you can create a personalized coverage map for the carriers you travel with and the minimum coverage type (LTE, 4G, 3G, 2G, roaming) you seek.

The maps are stored on device after downloading the app, so you don't need to have coverage to find out which direction to head when you can't find a signal. It's also useful for planning your next stop.

Crowdsourced Coverage Maps

Of course, just because a carrier claims they have coverage, that doesn't mean you'll be able to find it. There are some wonderful resources out there that aggregate crowdsourced signal and speed reports and create a coverage map based on what other customers are actually seeing in certain locations.

Give these free websites and apps a try:

- **Sensorly** - http://www.sensorly.com and an iOS and Android app of the same name.

- **OpenSignal** - http://www.opensignal.com and an iOS and Android app of the same name.

- **RootMetrics** - http://www.rootmetrics.com/us/ and an iOS and Android app called 'CoverageMap'.

Of course, with crowdsourcing, the maps are only as useful as the data they collect from users of their apps. These resources tend to have good data for urban areas where they have a strong user base. But, when you get to smaller cities, the maps can show no coverage at all.

This discrepancy is why we created *Coverage?* - which will have a better guess at where coverage might be based on carrier maps.

Campground Reviews

You'll frequently find reports of cellular coverage and Wi-Fi performance hidden within, or better yet, listed as a review field in campground reviews. Here are some of our favorites:

- Campendium.com – This review site includes specific fields for reporting coverage on each of the major carriers.

- Freecampsites.net – A database of remote camping and boondocking options with a field for reviewers to report their cellular signal for each carrier.

- RVParkReviews.com – This is one of the longest running review sites, and fellow nomads tend to leave coverage reports within their reviews.

- RVParky.com – Another popular review site where you might find coverage mentions within reviews.

- AllStays.com – Their app and website also have reviews where past campers might have noted cellular coverage.

Marina & Anchorage Reviews

And for those exploring by water:

- ActiveCaptain.com – It's not unusual to see reports of internet connectivity within reviews, though pay attention to the posting date; many reports may be years out of date.

And of course, don't just read reviews - leave your own reviews to help other travelers plan too!

Testing Your Connection

As you've learned so far in this chapter - wireless data is finicky, and many things can impact the signal. Bars and signal strength are only small indicator of the actual data performance you'll get.

If you are going to invest any effort in improving your mobile data connection, you need to have ways to measure the impacts of your changes.

Focusing on more bars is not enough.

To really evaluate a mobile network connection, you need to keep a close eye on your actual speed, latency and reliability testing results.

Speed Testing Services

There are numerous speed testing services and apps. These are the ones we regularly use:

- Ookla Speedtest (www.speedtest.net)

- Ookla Speedtest App (www.speedtest.net/mobile/) - For iPhone, iPad, Android, and Windows phones.

- Speed Of Me (www.SpeedOf.Me)

- Fast.com - Netflix's performance test focused exclusively on video streaming speeds. Useful for detecting if your video streams are being throttled.

If you ever get results that seem odd, try another service.

Understanding Speed Test Results

The results you get from a speed testing app might include:

• **Latency (aka Ping):** This is the time in milliseconds it takes for a request from your computer to reach the speed-test server and to return - like the ping of a ship's sonar. The higher the number, the slower the speed.

Latencies under 100ms are good and under 50ms are great. Latencies over 500ms (half a second) feel painful.

This measurement is particularly important for online gaming, but any interactive task can begin to suffer with higher latencies. For general surfing, it will be noticeable as that blank pause when you first request a new website.

• **Jitter:** Jitter is an indication of how much the connection latency varies moment-to-moment.

• **Packet Loss:** Some apps report a percentage of "packet loss." Think of this as letters lost in the mail. Substantial and persistent packet loss on a connection means the connection is unreliable, and web pages may not reliably load.

• **Download Speed:** Reported in either kilobits per second (Kbps) or megabits (equivalent to 1000 kilobits) per second (Mbps). This is a measurement of the maximum speed that data is able to flow to you from the speed-testing server. Speeds over 5Mbps give a good surfing experience and are suitable for most HD video streaming. Over 20Mbps will feel awesome and required for 4K HD video streaming. Speeds under 1Mbps start to make the modern internet feel slow, and speeds under 500Kbps are downright painful.

Download speeds have a particularly large impact on streaming audio and video. If the speeds aren't able to keep up with the resolution you've selected, you will experience stuttering, pauses and long buffering delays. And of course, the slower the speeds the longer it will take to download large files.

• **Upload Speed:** The opposite of download speed, the upload speed tells you how fast data is able to get from you to the speed-test server. Upload speeds are almost always substantially lower than download speeds. For many typical internet tasks upload speeds don't have a huge impact. But, upload speeds are critical for two-way video chatting, video broadcasts and uploading large files like photos, videos or cloud-synced backups. Speeds over 500Kbps are the bare minimum for low resolution video chat. Speeds over 1.5Mbps should deliver smoother results at higher resolutions.

Speed Test Tips & Tricks

Speed tests can vary a lot from moment to moment, and a lot of that variability may not have anything to do with your personal network connection.

To get a sense for the actual health of your connection you can run several speed tests over the course of 10 or 15 minutes. Doing this can help you get a better sense of what average speeds you are actually achieving.

Most speed-testing sites and apps have a way to change the test server, letting you select a different server to communicate with and test against. Trying different servers can help you rule out whether strange results are isolated or not.

If you are comparing usage between two devices, make sure that your speed tests are using the same server! An overloaded server can make one connection test out slower than another, when in fact it might actually be faster.

And finally - keep an eye on your data usage. Speed tests work by sending large chunks of data back and forth to the server, timing how long it takes. Excessive speed testing can burn through your monthly data bucket rapidly if you are not careful.

There's Only So Much You Can Do

A lot of people invest big dollars into antennas, cellular boosters, and Wi-Fi gear expecting miracles. And in the end, many end up being disappointed when their lofty expectations are not met.

Antennas and boosters can only do so much. If there's nothing to boost, an antenna or booster will be of no help. No amount of signal amplification can make something out of nothing.

Even if you get a stronger signal thanks to a booster, if the real speed bottleneck is located upstream, then you might not actually see any practical improvement anyway, at least when it comes to your online experience.

If you keep realistic expectations, however, a cellular booster and long-range Wi-Fi gear can grow to become some of your most essential tools.

When they work, they becomes absolutely indispensable.

Cellular Data: Carriers & Plans

Cellular data is probably the easiest and most accessible option for getting online in most places across the USA.

There have been amazing advancements in speed, coverage, and reliability over recent years and cellular technology continues to advance.

But as simple as it can seem on paper, cellular is also sometimes a confusing subject, primarily because there are just so many options!

When considering cellular, you have to choose which carrier(s), which plan(s), what equipment, and how much speed and data you actually need.

The Four Nationwide Carriers

The first choice to make is which cellular carrier, or carriers, to set up service with to best cover your mobile data needs.

In the US, the four major nationwide carriers are (ranked by size):

- **Verizon**
- **AT&T**
- **T-Mobile**
- **Sprint**

Quick Glance

Pros

Widely available

Easily accessible

Can be blazingly fast

Cons

Expensive

Limitations

Variable signals

Cellular Data: Carriers & Plans

All four have embraced the same underlying fourth-generation (4G) cellular network technology, known as LTE. But they all have very different legacy 2G and 3G networks, coverage maps, compatible devices, supported frequency bands, and expansion plans going forward towards 5G.

If you were living in one city or neighborhood, you could ask friends for their experiences with their carriers to determine which would serve you best in that particular local area.

But as a traveler, you will be moving around a lot. And, in different locations, different carriers excel.

You need to pick carriers that are well suited for all the places you plan to go, as you'll be dependent on where your carrier has built towers and/or has roaming partnership agreements. No network has coverage everywhere, and even if you bought a plan and device from each of the major carriers, *you would still* encounter times you just can't find a signal.

And you'll also have to select the carriers that have data plans available that will best meet your connectivity needs and budget.

> **Bottom Line:** There is no singular best carrier for every nomadic traveler!

A Note about Regional & Local Carriers

In addition to the 'big four' national carriers, there are a number of smaller regional and even local carriers that own and operate their own networks.

Some of the more prominent examples of these include U.S. Cellular, C-Spire Wireless, nTelos, Cellcom, and Cellular One.

These smaller regional carriers are usually poor choices for travelers, unless you know that you are primarily going to be spending time in areas where they have a strong native presence.

Even if the regional carrier has nationwide coverage through roaming agreements, if you're utilizing the service primarily outside its home region, you can find yourself running into all sorts of restrictions and limitations. All the way up to having your service summarily terminated!

Overview of the Four Main US Carriers:

Verizon

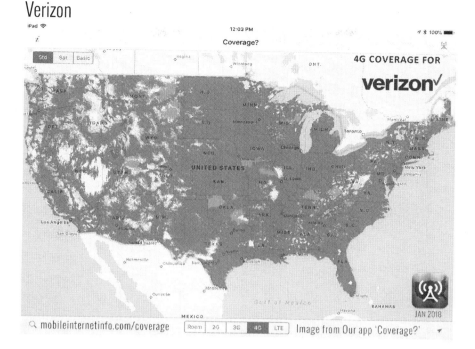

Verizon is the largest cellular carrier in the USA. Verizon has the widest overall coverage area, the most deployed LTE, and typically good overall performance.

For these reasons, if you're only going to choose one network, Verizon seems like the natural top choice.

But Verizon's plans can be pricey and riddled with limitations that keep them from being attractive as a primary mobile internet source.

And because Verizon's network is known to have the widest coverage and is the most popular network amongst nomads, it's actually not uncommon to pull into some popular areas to find the local Verizon tower overloaded and sluggish during peak times.

Verizon Device Tips:

For maximum compatibility, make sure that your Verizon devices support LTE Bands 2, 4, 5, and 13 - and that they have support for Verizon's legacy CDMA 3G network too.

Verizon's next cellular expansion will be on LTE Band 66, so keep an eye out for support for this band too.

AT&T

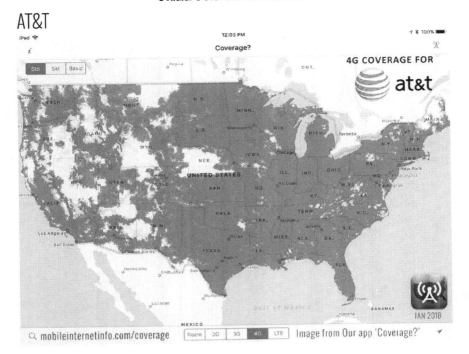

mobileinternetinfo.com/coverage | Roam | 2G | 3G | 4G | LTE | Image from Our app 'Coverage?'

AT&T is the second-largest carrier and is a formidable rival and a great complement to Verizon for nomads.

AT&T's LTE network often lags Verizon in coverage and speed, but there are many parts of the country where AT&T excels.

Up until 2017, AT&T's data plans weren't overly attractive as a home internet replacement. With their newer plans AT&T has become a more common top choice as a primary carrier for nomads.

A combination of Verizon and AT&T on board gives the widest coverage across the country.

AT&T Device Tips:

For maximum compatibility, make sure that your AT&T devices support LTE Bands 2, 4, 5, 12/17 and now LTE Bands 29 and 30 too.

AT&T is calling its most advanced LTE technologies "5G Evolution", so for the best possible future compatibility and performance keep an eye out for devices AT&T labels as compatible.

AT&T's legacy "4G" HSPA+ network has great performance and speeds nationwide, and occasionally can even outperform AT&T LTE.

T-Mobile

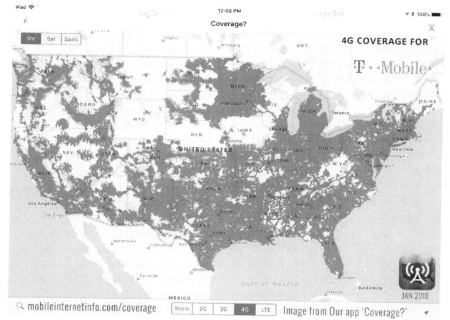

mobileinternetinfo.com/coverage   Roam 2G 3G 4G LTE   Image from Our app 'Coverage?'

T-Mobile has been the carrier to watch, blowing past Sprint to take a solid third place in nationwide coverage and soon T-Mobile may even rival Verizon.

T-Mobile's biggest achilles heel has been its lack of raw coverage, particularly in rural areas and indoors. When T-Mobile does have coverage, its network speeds are consistently some of the fastest.

T-Mobile is, however, moving aggressively to fill in its coverage gaps. In 2017 T-Mobile acquired a huge chunk of 600MHz cellular spectrum, and in the years ahead will be deploying more enhanced LTE service nationwide.

But to benefit you will need a compatible device capable of taking advantage of this new band.

T-Mobile Device Tips:

For maximum compatibility, make sure that your T-Mobile devices support LTE Bands 2, 4, and 12 - and support for the new LTE Band 71 will be required to take advantage of the 600MHz expansion.

As of the end of 2017, LTE Band 71 support is nearly non-existent in cellular devices, but more options should be available in the future. T-Mobile customers should consider putting off device upgrades until Band 71 compatible options are available.

Sprint

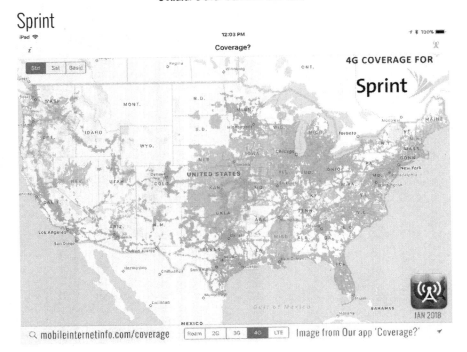

Sprint's biggest advantage is that it tends to have affordable plan options, especially with the promotional plans offered to lure new customers, or the special deals available only through third party resellers.

But the biggest downside of Sprint for nomads is its extremely limited nationwide coverage map.

The vast bulk of Sprint's usable fast data coverage is only found in core urban areas and along major interstates. Outside of that you're often roaming with very slow speeds - if you can get online at all.

If you're planning to stick to urban areas, Sprint might be worthwhile. But for most nomads, it is at best, a backup option.

Sprint Device Tips:

For maximum compatibility, make sure that your Sprint devices support LTE Bands 25, 26, and 41.

Some Sprint devices support a technology called "HPUE" which allows for extended range cellular connections on Sprint's fastest band, LTE Band 41. HPUE support is still rare, but is worth seeking out to get the most out of Sprint's network.

For more on the carriers and their plans:

The carriers are constantly changing up their offerings and features to stay competitive with each other - and expanding their coverage maps.

Before choosing your carriers, be sure to check the Mobile Internet Resource Center for our current take on each.

Our Guide to the Four Carriers, and Current Top Pick Plans:
www.MobileInternetInfo.com/four-carriers

Our Cellular Plan & Pricing Guide:
www.MobileInternetInfo.com/dataplans

On-Device Data vs. Hotspot / Tethering Data

The data included with a cellular data plan is usually expressed as the data that is available for on-device usage. That is, how much data you can use when doing things online directly on your smartphone or cellular enabled tablet.

A separate feature of a cellular data plan is often the one that matters most for travelers trying to replace a home internet connection is personal mobile hotspot or tethering.

Mobile hotspot / tethering is also the feature that tends to have the widest variance in restrictions from plan to plan.

Creating a mobile hotspot / tethering is what allows you to use your cellular device to create your own personal network out of your cellular data allotment to get laptops, streaming devices, gaming systems and routers online. For travelers, it is the feature that allows cellular data to become a mobile internet solution.

Many phones and tablets today, especially smartphones, can be used as modems for computers. This allows you to share your phone's data plan with your other devices either wirelessly or via USB. There are dedicated hotspot devices that serve this function too.

You'll see the terms "tethering" and "hotspot" tossed about interchangeably, which isn't too important, as both actions are treated the same by carriers in terms of usage. Here is what they typically mean:

- **Tethering:** This is the term used when you are using a USB cable to connect your cellular device to your router or computer, sharing the connection. As a bonus, tethering typically keeps your device charged up while sharing!

- **Personal or Mobile Hotspot:** This refers to when you turn your cellular device into a Wi-Fi router creating a hotspot without a tangible cable. This allows your cellular device to share its internet connection with multiple nearby devices and computers wirelessly.

Most stationary cellular consumers also have a home internet connection (cable, DSL, satellite) they use for the majority of their online needs. The mobile hotspot feature is there for when they are away from home to get their laptop online for a little bit without having to use public Wi-Fi sources.

The carriers don't yet have the capacity to serve the role of being everyone's home internet replacement, which is why there are restrictions on mobile hotspot use.

Some cellular plans prohibit sharing any of your data with other devices at all, or only allow it at reduced speeds. Others place limits on how much shared data is available at high speeds before being throttled for the rest of the month.

Most postpaid plans now include at least some allowance for hotspot use.

This is something you need to pay particular attention to when shopping for a plan. Don't anticipate that the sales rep at your carrier's store is going to be well versed in the nuances of their plans involving this feature. It's just not something that matters to most of their customers.

Understanding the Limits of "Unlimited" Data Plans

Starting in 2017, all of the major carriers re-introduced unlimited data plans. Some carriers also have grandfathered legacy plans that might still be available if you know where to look. Other resellers have even managed to find ways to offer unlimited data plans to consumers too.

Not all unlimited plans are created equally, however. You'd think "unlimited" always means no limits - but it doesn't. These days it basically just means no automatic overage charges.

Spectrum, or the amount of bandwidth available for the carriers to deliver services, is a limited resource that still needs to be managed. As the demand for data increases, the term "unlimited" has been redefined to try to find balance between capacity and consumer desire.

Understanding these restrictions is the key to determining if an unlimited data plan will meet your needs, especially if you're seeking a home internet replacement on the road.

The most typical restrictions on unlimited plans are slowing down the speeds of the data delivered. Sometimes, unlimited high speed data is only provided for certain activities.

Slowing Speeds Down

Slowing down speeds can get a bit confusing, and you'll see terms like Throttling and Network Management mentioned in fine print. Here are the differences:

Throttling: Is when a carrier or reseller **always** slows down speeds. This might be for the entire month, only after a certain amount of data is used, or for specific activities.

Network Management: Is when a carrier reserves the right to de-prioritize the traffic of high bandwidth users. The carriers generally have a set threshold before this kicks in (i.e. 22GB of usage in a month). After that threshold is reached, if the customer is on a cellular tower that is experiencing high usage, the connection **might be temporality slowed down**. Once the tower is no longer congested, full speeds resume. When on underused towers, you may never even experience network management (meaning full speed ahead!).

Example: Imagine you are driving on a highway that has a high occupancy / car pool lane. During normal traffic, all lanes are at full speed because there's not much traffic. But, during rush hour, if you qualify to use the car pool lane, it's beneficial to avoid the bumper to bumper traffic in the 'normal lanes.' Network management is like this - after you hit your threshold, you can no longer use the carpool lane. But depending on the traffic, you may never even notice.

On Device vs Mobile Hotspot Use

As we discussed above, the carriers don't want to become a home internet replacement for everyone, so unlimited data plans usually come with restrictions on mobile hotspot use. Pay particular attention to these restrictions.

If the carrier offers a dedicated mobile hotspot device to add on to an unlimited plan (such as a Jetpack/MiFi) , the mobile hotspot restriction may or may not apply to that line. Make sure you read and understand the restrictions before purchasing a device and plan.

Additionally, on cellular plans, any video streaming allowed by a carrier is intended to be used on a smartphone or tablet. Trying to connect a streaming device (like Apple TV, Chromecast, etc) via mobile hotspot will count against mobile hotspot caps.

See our TV & Streaming section in the **Use Cases** chapter for ways around this.

Video Throttling

Streaming video is one of the most data intensive things you can do on the internet. But it's something customers are demanding as more and more entertainment moves online.

As such, the carriers have been trying to capitalize on the customer demand by offering "unlimited" streaming options, while at the same time retaining network viability by artificially capping video stream speeds to limit video playback resolution.

Shopping for Unlimited Plans

Beyond the direct unlimited data plans offered by the carriers, there are often 3rd party reseller plans available that may not be as restrictive.

There may also be legacy unlimited plans available that were created before restrictions were common. A popular one for years has been truly unlimited Verizon Grandfathered Unlimited Plans, which at the time of this publishing are still indirectly available.

The options are constantly changing and we keep a close eye on it.

Our guide to unlimited data plans, including all of the current options, restrictions and current top pick plans:
www.MobileInternetInfo.com/unlimited

MVNOs, Resellers & Prepaid Plans

Mobile Virtual Network Operators (aka MVNOs) do not own their own cellular networks. Instead, they buy service in bulk from the major carriers and resell it under their own brands.

You might recognize some of the more popular MVNOs such as StraightTalk, US Mobile, FreedomPOP, Project Fi and Consumer Cellular.

Via an MVNO or reseller you can get better pricing, no contracts and no credit checks. But these deals often come with reduced features including slower speeds, limited video streaming, no roaming, or mobile hotspot restrictions.

Usually, an MVNO can't advertise who is providing the underlying network, but if you explicitly ask or do some online digging, it isn't hard to figure out.

Some brands are actually owned and operated by the big carriers themselves (for example - Cricket Wireless by AT&T or MetroPCS by T-Mobile). This allows the carrier to target niche markets without diluting their national brand.

Most carriers also offer prepaid or pay-as-you-go plans that allow customers to only pay for service when they need it, rather than making month to month commitments. This makes setting up service super easy.

The Downside of MVNOs & Resellers

The biggest downside of MVNOs is that the offerings may not last and terms can change with little notice.

Every so often a reseller comes along with an offer that seems too good to be true. Frequently these offers do not last for long once the underlying network provider starts to object or realizes their costs to deliver the service are higher than anticipated.

We've tracked many resellers over the years that have come and gone.

The important lesson here is to be ready to jump on great offers when they come along, but remain aware of the risks.

Always have a backup plan in place in case service gets cut off suddenly.

And keep in mind that the "no contract" terms go both ways!

Always resist temptation to break existing contracts or give up irreplaceable grandfathered plans for a reseller's offer which may not last.

Shopping for Plans

Each carrier and MVNO will have its own set of plans, ranging from prepaid plans that are ideal for short-term needs or those who want more flexibility, to post paid plans that generally require passing a credit check.

Here are some additional considerations to keep in mind when looking at all of the various plan options out there.

Contracts & Equipment Installment Plans

None of the major carriers require contracts anymore. Instead, many have shifted to equipment payment plans that separate out the financing of your device from the service plan itself. This still technically ties you to the carrier until the device is paid off.

Since the option for a contract still exists, make sure you know if you enter one and what the penalty is for breaking it or suspending it. If you're getting a device for free or at a discounted price, there's probably a contract with early termination fees involved.

If you want to avoid contracts or payment plans, purchase your device at full price or bring your own by purchasing used equipment.

For travelers, we generally recommend minimizing the number of long-term contracts for service. You may find your travel plans change, you might take an extended international excursion, or go on nomadic hiatus. Or maybe a better plan on a different carrier comes along that renders your current plan unattractive.

But, sometimes securing a fabulous limited time offer is worth the downsides of contracts to lock the plan in longer term.

Vacation Mode

If you find yourself staying in a spot for a while where you don't get enough signal to keep online, or where you might have access to other reliable options, you might want to temporarily suspend service on some of your plans.

Investigate upfront if that's an option with your plan and if there are time limitations or fees involved.

Re-Shop Plans Regularly

The carriers are constantly changing their plans to remain competitive in relation to each other. It's very worthwhile to reconsider your plans at least annually, as your carrier may not automatically adjust your pricing or features as they roll out better plans.

Just like regularly checking the oil in your engine, you should also make checking your plans a regular part of your mobile maintenance schedule.

Understanding Roaming

Roaming is when a cellular carrier has agreements with other networks to utilize their towers, helping the carrier provide connectivity to their customers who are just passing through areas they don't directly service themselves.

Roaming is essential to regional carriers and the smaller national carriers, since they lack the vast networks of AT&T and Verizon. But even AT&T and Verizon use roaming behind the scenes to help flesh out their maps.

While carriers rarely charge you extra for domestic roaming, they do tend to have special data roaming limits to keep your usage from costing them. Or their roaming terms may vary by the type of partnerships they've entered into - such as some roaming being included as if its native coverage.

Most carriers have roaming restrictions when traveling internationally as well, or even extra fees.

Most of the carriers will send you a text message or email to let you know you've hit your roaming cap, although some will just flat out terminate you for exceeding the limits.

Knowing When You Are Roaming

On most mobile devices, right next to where the signal strength you are getting is displayed, you can see the name of the carrier you are connected to. Usually this is set to your provider (i.e. AT&T or Verizon).

But, when you're roaming, you might see another carrier's name listed instead, letting you know that you are in a roaming area.

By default, some carriers and devices don't switch the carrier's name displayed until **after** you've hit your roaming cap. If you're planning to be traveling a lot and want to be better able to tell when you are roaming, you may want to call your carrier and ask if they can make that switch now.

We actually were first inspired to create our *Coverage?* app after running into roaming limits. With the app, you can quickly check if you're in a roaming area before preparing for settling in for an extended stay.

Roaming into Canada & Mexico

Most plans from the major carriers now include free roaming into Canada and Mexico, a fabulous feature for travelers both over and near the border.

But many legacy plans, MVNOs and smaller carriers do NOT include free roaming into Canada and/or Mexico, and any inadvertent usage will be billed as very expensive international data.

Know your plan limits, and be careful near borders. Even if you do not cross them, wireless signals can!

Our Guide to Keeping Connected In Mexico:
www.MobileInternetInfo.com/Mexico

Our Guide to Keeping Connected In Canada:
www.MobileInternetInfo.com/Canada

Cellular Data: Gear

Once you decide on the carrier(s) you want in your arsenal, you have to decide what specific equipment makes the most sense for getting online, optimizing your signal and sharing the connection.

The basic options include dedicated devices that are restricted to data only (such as mobile hotspots and modems), or putting cellular tablets and phones to work serving double duty providing an internet connection for your other devices.

What cellular connectivity hardware makes the most sense for you depends on how many devices you're trying to keep connected, how many people you need to provide internet for, how many different carriers and data plans you want to keep active, where you think you'll need internet access, and what devices you already own.

Regardless of what gear you decide on, for maximum coverage and speed we recommend investing in the latest technology. Purchase the most recent cellular devices you can, and plan on upgrading some of your hardware as often as every year or two to stay current.

The options change frequently. When it comes time to shop for your specific equipment, you can check our latest reviews and comparative guides.

We also keep this chapter updated online at:

Smartphone, Hotspot or Routers Guide:
www.MobileInternetInfo.com/mifi

Smartphone / Tablet Hotspotting & Tethering

The simplest way to share a cellular data plan is by using the built-in mobile hotspot or tethering feature provided by most smartphones and tablets.

It's a feature supported by most cellular devices that can be turned on in the settings menu.

Whether you are using USB or creating your own Wi-Fi, your local connection is private and is safe from snooping.

Advantages:

- On most data plans direct with the major carriers, mobile hotspot support tends to be included at no extra cost. But be sure to check the exact terms of your plan as to what is included - and how much data can be used via hotspot.

- No extra devices to manage - sweet and simple.

Disadvantages:

- Not ideal for households with multiple people. What happens if the person with the hotspot-enabled phone heads out to run errands?

- Many devices go to sleep when there's no ongoing connection activity, so you may need to wake your device up and restart the hotspot feature after every period of inactivity online.

- Hotspotting from a smartphone or device drains its battery pretty quickly. Make sure the device is plugged in.

Recommended For:

Solo travelers, those not dependent on internet for critical tasks, for access to a secondary cellular network (ie. if your primary is Verizon with a MiFi, perhaps you access AT&T when needed from a smartphone) or for out and about internet access.

For our Tips for Selecting Smartphones:
www.MobileInternetInfo.com/smartphones

Mobile Hotspots & Modems

Also sometimes referred to as a 'Jetpack' or 'MiFi', mobile hotspots are small self-contained units that receive a cellular data signal and then create a private local Wi-Fi network using your cellular data plan.

A mobile hotspot device is a cellular modem and a basic Wi-Fi router combined into one small consumer grade unit.

Most mobile hotspot devices tend to be able to support 5 to 15 connected devices at once, and many can also USB tether to a laptop or mobile router. They commonly have a battery built into them, which allows you to take your service with you on the go.

Advantages:

- These can be a fairly easy plug-and-play solution ideal for users who don't want to have to learn to manage other more complex options. Just turn it on and connect.

- It's a dedicated device that can be left in your tech cabinet, plugged in, sitting next to a cellular booster antenna, and mostly forgotten about. Many people keep hotspots running 24/7.

- Some hotspots have antenna ports, allowing you to directly connect external antennas for signal enhancing - giving you an extra option for better performance.

Disadvantages:

- Hotspots are not full-featured routers, and typically do not support ethernet ports, WiFi-as-WAN connections, or other advanced router features.

- Hotspot Wi-Fi range will usually cover most indoor areas of typically sized RVs and boats, but range is limited beyond that.

- Check your cellular data plan carefully. Some carriers will allow these devices to be added to an "unlimited" data plan, but actually place restrictive hotspot data caps on the plan.

Recommended For:

Multi-person or multi-device households, those who depend on cellular data for critical tasks, those who want access to the latest network technology, those who don't want to fiddle with tethering from a smartphone, and those who need their home to have internet access even while they're away (such as remotely checking in on home automation systems).

Mobile Hotspot Variants

Aside from self contained battery operated hotspot device, there are some other variants you might encounter when looking at your options.

Connected Car Devices

These types of devices are built into some vehicles right from the factory as part of an OnStar (or similar) package, or a small cartridge plugged into the OBD-2 diagnostic port found under the dash of other vehicles.

Some Connected Car devices can create a Wi-Fi hotspot whenever the ignition is engaged in addition to providing vehicle diagnostics.

The downside of Connected Car devices is that most will only work while the ignition is engaged, making them great on the go but of limited use while parked.

Traditional standalone mobile hotspots are usually more practical and technologically advanced.

USB Modems

Unlike a mobile hotspot, these devices do not create a Wi-Fi hotspot or allow multiple devices to connect at once.

Instead they provide a cellular connection for the single computer or compatible cellular-aware router that they are plugged into.

If you're traveling solo and just need to keep a single laptop online, a USB modem may be a simple and elegant solution for staying online.

Some laptops even include USB modems available as a built-in option.

But in a world of diverse connected devices, basic USB modems focused on traditional laptops are becoming increasingly rare.

Cellular Modem & Hotspot Guide:
www.MobileInternetInfo.com/lte-modems

Cellular Integrated and Mobile Routers

If you have more demanding needs than even a mobile hotspot can provide, the next step up involves incorporating a full-featured mobile router.

Among many other features, routers tend to provide ethernet ports for wired networking, support for more devices sharing the Wi-Fi network, stronger Wi-Fi broadcast range and can allow for easier switching between different upstream connection options.

Some mobile routers use USB cellular modems or tether to basic mobile hotspots to get a cellular connection to share, and some higher-end models actually have cellular modems integrated right inside them.

But do be aware - most typical off-the-shelf residential routers will NOT work well for sharing a cellular connection in a mobile set-up.

To understand more about the features of mobile routers, and the different types of options available, see the **Routers** chapter.

Advantages:

- Makes it easier to set up a local wired and/or wireless network - with support for printers, streaming devices, gaming systems, network storage and back-up drives, and more.

- By funneling all of your connectivity options through a router, it is easy to change between cellular carriers, or to switch between cellular and WiFi-as-WAN (an outside Wi-Fi source - i.e. campground Wi-Fi) upstream connections.

- A router can usually create a faster and longer-range private Wi-Fi network well beyond what a smartphone or MiFi alone can.

Disadvantages:

- Cellular-integrated routers tend to be updated infrequently, so they often lag a year (or more) behind mobile hotspots and smartphones in supporting the latest cellular technologies.

- Routers introduce substantial complexity, require more networking knowledge to install and manage, and cost more initially.

Recommended For:

Those with multiple devices to keep connected, particularly those who want to easily switch between upstream connections.

Understanding SIM Cards

Almost all LTE compatible devices have a tiny removable sliver of plastic called a SIM, short for Subscriber Identity Module.

This little chip is what ties your phone number and plan with your device.

This card is often found lurking underneath the battery of many phones or hotspot devices, or in a tiny ejectable tray that can be removed by pushing a pin or paperclip carefully into a pinhole.

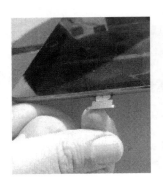

If you want to use your cellular plan in another compatible device, it's usually as simple as moving the SIM card to the new device.

However, not all plans are compatible with all devices, so do your homework first.

Some reasons you might want to do this:

• You can protect your high end phone by moving the SIM to an old beater phone when you are heading out into rough conditions.

• You can get a SIM (and a local phone number) from a local cellular company when traveling.

• You can upgrade your device without involving contacting your carrier or risking changing your current plan.

SIM Sizes

The SIM card sizes currently used are: Mini-SIM, Micro-SIM, and Nano-SIM.

These cards are all electrically identical, so it is actually possible to cut down a Mini-SIM to put it into a device that has a Nano-SIM slot. And, you can use a small plastic cradle to put a smaller Nano-SIM into a Mini or Micro slot.

You can buy a SIM cutter tool to make the process easy. But, a lot of potential headaches can be avoided if you make sure the devices you want to swap between all use the same size.

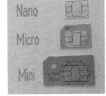

Locked & Unlocked Cellular Devices

SIM cards make it easy to move service between devices. If your device is compatible with the underlying frequencies and cellular standards, by swapping SIM cards you can even change carriers.

But only if your device is unlocked.

Carriers very commonly lock new phones and hotspots so that they will only work on that carrier's network. If you put in a SIM from a competitor, it will not work.

Verizon is the only major carrier that currently has a policy of selling all phones, Jetpacks, and tablets on postpaid plans unlocked - a very nice perk.

All of the carriers, however, are required to unlock your phone at your request once you have fulfilled any contract or device installment plan obligations. Prepaid devices can usually be unlocked too, however they tend to need to be in paid service for several months before they are released to be unlocked.

Make sure you handle getting devices unlocked BEFORE you head out on an international trip or try to gift your old phone to a friend.

Cellular Frequencies & LTE-Advanced Technology

All of the big four national carriers have settled on the same 4G/LTE networking standard. But, despite the shared standard, it is common for cellular phones and devices to be optimized to work with just a single carrier.

Multi-carrier devices are still rare, though growing less so.

This is because all four national carriers are building their networks on different frequency bands, and they also have older incompatible legacy networks to support for 2G, 3G, and voice services.

There are over a dozen defined LTE bands in active use in the United States, with more being added.

Every carrier owns multiple blocks of spectrum within various combinations of these bands. Not all phones and data devices actually have radios capable of tuning in to every frequency and communications standard.

Ensuring you have access to a carrier's entire network is necessary for the fastest possible speeds and the best possible coverage. There is some cross over of bands between the carriers, which is why a device optimized for one carrier may work on another carrier, but not optimally.

Newer LTE devices support LTE-Advanced technologies that actually allow signals on multiple LTE bands to be combined using a technique called **Carrier Aggregation**, enabling a huge potential performance boost.

Look specifically for this feature on any device you purchase, as it can be a night and day difference for mobile travelers who are constantly encountering multiple frequency bands on their travels.

Carpool Lanes in The Sky

A highway comparison works for helping to understand LTE bands. Think of LTE bands and cellular spectrum like lanes on a highway.

The more lanes a carrier can offer, the more simultaneous users can be supported, and the faster each user can go.

Older devices without support for the latest LTE network bands are limited to the most congested lanes on the road, while newer devices can often zip past in the carpool lane leaving older devices stuck in traffic.

This is one of the reasons why it is important to keep your connectivity arsenal up to date as your carriers continue to expand their coverage onto newer frequency bands.

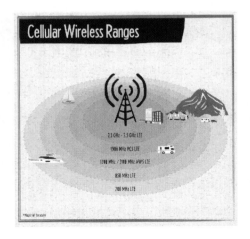

Each band is a short hand for the frequency of the wireless signal used. Lower frequencies tend to provide more range, and higher frequencies tend to have less range, but can provide more speed and capacity.

Cellular Data: Gear

Here are the current cellular frequencies & bands in use in the United States:

Frequency	Common Name	Band	Who Is Using It?
600MHz	UHF TV Channels 38-51	LTE Band 71	T-Mobile's major area for rural expansion.
700MHz (Lower)	700MHz Auction Block A/B/C	LTE Band 12	T-Mobile "Extended Range LTE"
700MHz (Lower)	700MHz Auction Block B/C (Subset of band 12)	LTE Band 17	AT&T LTE
700MHz (Lower)	700MHz Auction Block D/E (Download Only)	LTE Band 29	AT&T LTE (New in 2016)
700MHz (Upper)	700MHz Auction Block C (Upper)	LTE Band 13	Verizon LTE
850MHz	Extended CLR (Cellular)	LTE Band 26	Sprint "Spark" LTE
850MHz	CLR (Cellular -the original!)	LTE Band 5	Verizon 3G, AT&T 3G, AT&T 4G
1700MHz/ 2100MHz	AWS (Advanced Wireless Service)	LTE Band 4	Verizon XLTE, AT&T LTE, T-Mobile 4G, T-Mobile LTE
1700MHz/ 2100MHz	AWS-3 (Advanced Wireless Service – 2014 Auction Expansion)	LTE Band 66	Auction winners: AT&T, Verizon, T-Mobile, Dish
1900MHz	PCS (Personal Communications Service)	LTE Band 2	Verizon 3G, LTE; AT&T 3G,4G, LTE; T-Mobile 2G,4G,LTE
1900MHz	Extended PCS	LTE Band 25	Sprint 3G, LTE
2.3GHz	WCS (Wireless Communication Service)	LTE Band 30	AT&T LTE (New in 2015)
2.5GHz	BRS (Broadband Radio Service)	LTE Band 41	Sprint "LTE Plus"
5GHz	LTE-U (Unlicensed) / LTE-LAA (License Assisted Access)	LTE Bands 252, 255	Experimental: Verizon, T-Mobile

Cellular Signal Optimization - Antennas & Boosters

One of the problems with cellular internet is that the signal can be quite variable depending on many factors – your device, the tower location, weather, how many people are also using the tower, local terrain, nearby buildings, and even your own vehicle's construction.

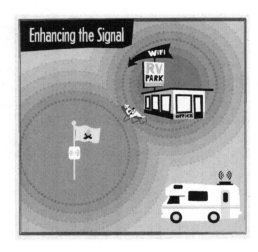

You can, however, buy various external antennas and cellular boosters that can improve the situation.

Sometimes these are lifesavers and can make a finicky signal usable enough to get your work done or stream a movie.

And other times they might make no difference at all, or might even degrade your cellular data performance.

Enhancing cellular signal and data performance is a tricky subject, and sometimes requires trial and error at each location for each type of device & cellular carrier.

It's a very in-depth topic with a lot of variables and background information to consider which is beyond the scope of this book.

For more on Cellular Signal & Data Optimization:
www.MobileInternetInfo.com/cellsignal

Here is an overview of the basic options for improving a weak cellular signal:

Placement

A phone in your pocket pressed up against a big dense bag of saltwater (you) is going to get a lot less signal than a tablet sitting on a desk, and both will be at a substantial disadvantage to a mobile hotspot placed in a window.

Sometimes just a slight relocation can make all the difference in the world. Don't be afraid to experiment.

Place your hotspot near a window or even get a suction cup soap caddy to mount your device in a window with better signal (but mind any direct sunlight that could cause overheating).

External Antennas

Very few phones and tablets have antenna ports anymore, but many mobile hotspots and USB modems do, and all commercial-grade cellular routers do.

If you have an antenna port, you can attach a more capable external antenna to that device for better reception. This allows you to place the antenna were it is most optimal - such as installed on your roof - and not worry about the placement of your cellular device.

More and more cellular devices now have two antenna ports. Installing a MIMO designed antenna or even a matched pair of antennas preserves antenna diversity and allows for LTE's MIMO capabilities.

Diversity provides for increased performance in noisy signal areas with a lot of signal reflections, and MIMO allows for multiple data streams to be combined for greatly enhanced speed.

To better understand what MIMO is and how it works – refer back to the **Going Wireless** chapter earlier in this book.

For more on antenna selection and installation:

For our guide to Cellular Antennas:
www.MobileInternetInfo.com/cellularantennas

For our guide to Antenna Installations:
www.MobileInternetInfo.com/antennainstall

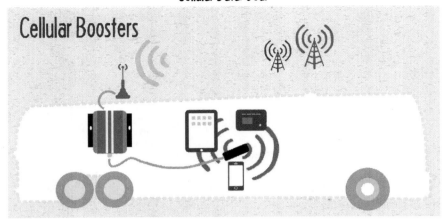

Cellular Boosters

A cellular booster picks up a signal with an external antenna, amplifies it, and wirelessly rebroadcasts it via an internal antenna to provide a stronger signal for all the nearby interior devices whether they have an antenna port or not.

A cellular booster provides several advantages to improve your connection, including increased power for upload capacity.

These devices can range from simple systems that just boost one mobile device at a time to systems that can boost multiple devices at once across multiple carrier networks.

Boosters are not miracle devices though. There has to be at least some signal present for them to work.

And in some situations, they may not be the best choice. MIMO antennas can often out perform boosters due to diversity, which boosters lack.

Boosters also require special installation considerations. Setting up in the smaller space of an RV or boat requires finding a delicate balance between antenna gain, amplifying power and avoiding signal feedback between exterior and interior antennas.

Sometimes, they can be the best option get aid in getting online. But boosters are a pricey investment that you should research more before deciding if they are right for you.

Our Guide to Mobile Cellular Boosters:
www.MobileInternetInfo.com/boosters

Bars vs. dBm vs Speed

Everyone knows that more bars is a good thing. You might not know that different phones and operating systems calculate how many signal bars to display very differently.

This means that comparing bars, unless you are on the same exact model and same carrier, is actually a very poor way to express signal quality.

In addition to raw signal strength, the phone may be measuring network congestion and other variables to calculate how many bars to display.

This can make it harder to measure the impact a booster or antenna is having, because improving the signal strength might not register as more bars right away, especially if the phone is focusing on network congestion.

To objectively compare signal strength, you need to look for the dBm reading, which may be hidden deep inside that advanced status menus on your device.

This readout will show the actual power in milliwatts being received by an antenna.

This is a logarithmic scale. A 0dBm reading represents a single milliwatt received, and every change by 10 up or down represents a 10x change in received signal power.

The power levels picked up by a cellular antenna are fractions of a milliwatt, so the dBm readout will be a negative number.

> **–50dBm** would be considered an awesome signal.
>
> **–60dBm** is 10x weaker, but still great.
>
> **–70dBm** is 100x weaker.
>
> **–80dBm** is 1,000x weaker.
>
> **–90dBm** is 10,000x weaker.
>
> **–100dBm** is 100,000x weaker – and is where boosters & antenna can have the greatest impact.
>
> **–110dBm** is a million times weaker than –50dBm and is usually barely usable.
>
> By the time you see **–120dBm**, the phone or hotspot has probably already given up and switched to "No Service."

Modern radios can work wonders with weak signals.

In the past, **-95dBm** would have been considered weak, but a modern LTE radio can work with signals of **-100dBm** all the way down to **-110dBm** and often beyond.

TIP: Focus on Speed, Not Signal

Because modern radios are so skilled with eking out maximal performance from weak signals, it is actually common for boosters to increase signal strength (and number of signal bars) while often actually having no impact on performance.

Sometimes a booster might actually even degrade performance because they don't utilize the all important LTE technology of MIMO.

It's also common for your signal enhancing attempts to switch which frequency bands you're connecting to - which can have a substantial difference in performance depending on the current load of the tower or how far away you are. As an example, 2 bars of signal on Band 4 might perform better than 5 bars on Band 12 at your current location.

When optimizing your signal, focus on performance testing to determine what impacts your changes are making, not just on the signal strength or the number of bars displayed.

After all, a fast one-bar signal is certainly preferable to a slow five-bar connection!

Refer back to the **Going Wireless** chapter for how to test your actual performance and understanding the results.

Wi-Fi: Using Hotspots

Often the fastest, cheapest, and easiest way to get online is to use public Wi-Fi networks, and these are often pretty easy to find.

Many libraries, coffee shops, RV parks, marinas, breweries (yay!), motels, municipal parks, and even fast food restaurants now offer free Wi-Fi. There are also plenty of paid Wi-Fi networks to be found.

Though Wi-Fi has the potential to be blazingly fast, some shared Wi-Fi networks can be horribly overloaded. A public Wi-Fi hotspot is highly dependent upon its upstream source of internet (cable, DSL, satellite, etc.) and how many people are sharing that connection.

Quick Glance

Pros

Widely available

Easily accessible

Frequently free

Cons

Variable quality

Frequently unreliable

Security concerns

Clearly, not all Wi-Fi is created equally.

In some cases, the upstream connection may actually be little better than old dial-up modems. In some remote places, the upstream connection may actually BE an old dial-up modem!

Unfortunately, in many situations, even though you may be able to get online via Wi-Fi it may not be worth the effort.

The other major limitation of Wi-Fi is range.

Wi-Fi: Using Hotspots

While working in a cafe or brewery can be enjoyable, you may prefer to be 'at home' surfing the internet. Most Wi-Fi hotspots fall off to unusably slow connections just a hundred feet away from the base station, and in some parks and marinas only the spots nearest to the front office can reliably connect via Wi-Fi.

But with the right long-range Wi-Fi gear (see the next chapter), you can often manage to connect to a base station substantially farther away than your unaided laptop or tablet alone ever could.

Realities of Campground & Marina Wi-Fi

Although you would think that a campground or marina that advertises "Free Wi-Fi!" as prominently as it does 50A power hook-ups would have worthwhile service, you may not find that to always be true.

Providing Wi-Fi that can serve hundreds of bandwidth-hungry travelers, especially spread out over several acres, is very expensive to set up. Few locations have the IT expertise on staff to properly upkeep such a network either.

Doing it right requires a substantial investment in routers and access points, as well as needing a pretty hefty internet backbone to tap into.

Generally, even if the Wi-Fi is managed professionally, it is common for free networks to be only good enough for checking email and doing some basic surfing. All most travelers are assumed to need.

And even well run Wi-Fi networks are often seriously overloaded.

A connection that might be decent during the day while everyone is off exploring, might feel worse than dial-up during prime time.

It can take just one or two people trying to stream a movie, video chatting with the grandkids, or downloading a huge file to bring the entire Wi-Fi network to a grinding halt.

Wi-Fi: Using Hotspots

Some locations have outsourced the chore of providing Wi-Fi to a provider, like Tengo Internet or onSpot, who manage the bandwidth and network. Sometimes they charge extra for service, cap how much data you can use, or limit how many devices you can connect at once.

The theory is that these limits help fairly spread out the available capacity, and setting up a paid network keeps freeloaders who might not even be customers from tapping into the network.

Professionally managed Wi-Fi connections should in theory end up being faster and more reliable than open and free. Unfortunately that's a theory that doesn't always manifest into reality.

It's best to look at campground and marina provided Wi-Fi as a bonus if it's usable, but not to plan around it for anything critical or bandwidth intensive.

You may find great Wi-Fi – fast and without data caps. But these networks tend to be the exception, not the rule.

Free Public Hotspots

A lot of businesses and other public areas provide free Wi-Fi access to their customers.

If you have the flexibility and/or desire to take your laptop with you, these hotspots can offer great connectivity. Some folks actually rely on this method as their primary internet, seeking out cafes and libraries all across their travels.

And, if you can get your RV or boat within range of a public hotspot, you might even be able to use them from your recliner - no field-trip required.

Places known for their free hotspots include libraries, laundromats, McDonald's, coffee shops, Panera Bread, Lowe's (yes, the hardware store), Walmart, some rest stops, motels (higher end hotels tend to have paid internet), breweries, restaurants, and so many more.

There are several apps and websites out there that track known hotspots. It's usually pretty easy to stumble into these places when you need them.

Wi-Fi: Using Hotspots

Keep in mind businesses optimize their Wi-Fi network to provide service within their facility for their customers, not to beam service to a parking lot for non-customers. You may only get a usable signal inside their building.

You might have to ask for the password when ordering your coffee.

Some places might limit how long they'll let you stay on their connection, as they do need their tables for customers just arriving. If asked to move on, be respectful and comply.

Please respect that provided Wi-Fi is generally a courtesy that costs money to provide. Do your best to be a customer and tip your server well for taking up a seat in their area for an extended amount of time.

And always remember, not all Wi-Fi is created equal and your speeds may vary from location to location. Before buying a meal or coffee, run a speed test to make sure the bandwidth offered is usable enough for your needs.

Paid Wi-Fi Networks

There are some widely deployed Wi-Fi networks available to paid customers.

The nationwide Cable WiFi initiative involves several regional cable companies, all of which allow customers to roam freely between connected Cable WiFi hotspots.

With over 300,000 Wi-Fi hotspots and growing, if you are in an area served by one of these companies you might be surprised to find that you can get access in some very unexpected places.

Comcast's Xfinity WiFi is taking things even further and now has millions of "xfinitywifi" hotspots. Comcast has accomplished this by turning business and home customers' cable modems into public Wi-Fi hotspots. In one fell swoop they are offering fast Wi-Fi over entire neighborhoods. Non-Xfinity customers can get complementary Wi-Fi sessions or can buy hourly, daily or weekly passes.

If you see "xfinitywifi" or "CableWiFi" as an available hotspot, you can try this out as an option. If you are in a Comcast area, the Xfinity iOS and Android apps will help you find areas that are covered.

Check if your stationary family members and friends are customers of one of the participating cable companies. You might be able to get their permission to use their login to access the network.

Wi-Fi: Using Hotspots

There are also services like Boingo, which has over one million hotspots around the world, including many Tengo locations found within campgrounds. For a monthly fee you can get unlimited access at all of the partnered locations.

Borrowing Bandwidth From Friends

One of the most reliable ways to access Wi-Fi is by borrowing a cup of bandwidth from friends and family as you travel.

You'll find most folks with fast home connections are more than happy to share their unlimited high-speed bandwidth to do things like OS updates, download videos, software updates, or do a massive data syncing to cloud storage services.

It's a great way to combine social time and getting digital chores done.

Wi-Fi Frequencies & Congestion Issues

Wi-Fi is defined by the IEEE 802.11 set of standards and the specifications for any Wi-Fi-compatible device should indicate which variants (indicated by letters appended to 802.11, such as 802.11b/g/n/ac) of the standard are supported. The specifications should also indicate the frequencies used.

When two Wi-Fi devices try to connect with each other, theoretically they should negotiate the fastest and most recent connection standard that they are both compatible with.

TIP: If you are manually configuring a Wi-Fi device, channels 1, 6, and 11 are the ones that do not overlap and interfere with each other.

Most older Wi-Fi devices operate on an unlicensed 2.4GHz frequency band that provides for only THREE fully distinct radio channels.

Bluetooth devices and most cordless phones (remember those?) also operate in this same unlicensed 2.4GHz frequency band further adding to the congestion. In an urban area or even a crowded campground or marina, those three channels can get awfully overloaded with signals.

Microwave ovens also emit radiation in the same 2.4GHz band. If your neighbor is making popcorn, it can potentially grind Wi-Fi speeds to a halt for everyone nearby.

Despite all the interference, it is amazing how well 2.4GHz Wi-Fi works.

5GHz Wi-Fi: An Uncrowded Expressway

Newer and more advanced Wi-Fi devices also support Wi-Fi channels located in the uncongested 5GHz frequency band, where 23+ non-overlapping channels are sitting, usually vacant. This is the express lane compared to the 2.4GHz gridlock; a quiet library compared to a raucous bar.

But to take advantage of 5GHz requires different antennas and equipment specifically designed to broadcast on these channels.

Range Considerations:

Lower frequencies travel further and are better at penetrating walls, so while 5GHz 802.11ac may be the ultimate high-speed local wireless network within your RV or boat, it is not well suited for talking to distant access points.

For the time being, 2.4GHz 802.11n usually remains the technology best suited to longer range Wi-Fi.

Purchasing Tips:

If you are buying a Wi-Fi router, look for 802.11n (or 802.11ac), explicit mention of 5GHz, and especially "simultaneous dual band" support, which means the router can connect to clients using both frequency bands at once.

If you are buying a laptop or tablet, to make sure that you are able to to take advantage of faster speeds in the future make sure that 5GHz 802.11n or 802.11ac is supported.

For more on Wi-Fi, see our guide:
www.MobileInternetInfo.com/wifi

Wi-Fi: Range Extending Gear

So you found a Wi-Fi hotspot to use, but you can't get the signal while sitting at your RV's dinette or in the salon.

How do you get online via Wi-Fi, without needing to spend your days sitting at Starbucks and nights at the marina's or RV park's office?

Wi-Fi was never intended to be used for long range networking, but it is often possible to push the limits of what Wi-Fi is capable of.

But first test to make sure that it will actually be worth the effort.

The Wi-Fi Worthiness Test

A lot of people invest a small fortune in long-range Wi-Fi hardware, only to report back disappointedly that it hardly made any difference.

In a lot of these cases there just wasn't any worthwhile service to work with in the first place. If the campground or marina has slow and unreliable Wi-Fi in the front office near the hotspot, no amount of technology will be able to make things any better further away.

Before you invest time and money in installing extra gear, find out if the hotspot you're trying to connect to is actually worth the effort.

Do this by taking your laptop, phone, or tablet up as close to the hotspot as you can manage and then run some speed tests. Try out some typical web surfing. Maybe even stream a video.

Wi-Fi: Range Extending Gear

If the experience is a good one, then using long range Wi-Fi extending gear may help.

But if not, save yourself some frustration and find another way online.

If the performance is bad up close, there is nothing further you can do to improve it, except perhaps complain to management.

Wi-Fi Range Capabilities

The range in which you can reach and utilize a Wi-Fi signal is highly dependent upon the capability of the antenna and radio in your device.

Though nearly every laptop and mobile gadget made now comes with Wi-Fi capabilities built in, most aren't engineered for distance.

And few devices make any provision for using an external Wi-Fi antenna, so other than balancing your laptop in a window, there is no way to increase your Wi-Fi range without adding another device to the mix.

Generally, there is a rough hierarchy of Wi-Fi range capability, from the shortest range gear to the longest:

1) **Phones, Tables & Laptops -** Most small gadgets (including phones, tablets, etc) have very minimal Wi-Fi range. Laptops typically can connect via Wi-Fi further away than phones.

2) **Wi-Fi Range Extenders -** These types of devices connect to a remote Wi-Fi network and then rebroadcast it, acting like a relay.

3) **Indoor WiFi-as-WAN Routers -** These special routers connect to a more remote network, and use that as the upstream data source for your local private network. Many WiFi-as-WAN routers have relatively high-power Wi-Fi radios.

4) **High-Power USB Wi-Fi Network Adapters -** These are external Wi-Fi radios that you can plug into a laptop via USB, sometimes greatly enhancing Wi-Fi range. These often come with a long cable to allow placement in a window, or even outdoors.

5) **Outdoor Antennas -** A Wi-Fi antenna mounted outside can get line of sight benefits, and increase your range.

6) **Outdoor CPE -** Pairing an antenna with the radio outside gets the best range performance with line of sight, reducing signal loss and increasing transmit power.

We'll explain some of the more advanced options below.

And for some additional tips on things that can impact Wi-Fi connections, see the earlier chapter on **Going Wireless**.

WiFi-as-WAN & Range Extending

Some Wi-Fi routers support a feature called "WiFi-as-WAN" which allows them to use another Wi-Fi network as the upstream connection. WAN stands for Wide Area Network, or in other words, the internet.

Since a WiFi-as-WAN router will usually have a stronger radio and better antennas than any of your laptops or tablets, WiFi-as-WAN lets devices on your Local Area Network (or LAN) share access to a Wi-Fi network further away than they could ever reach alone.

This is a rare feature in typical off-the-shelf home routers, but is common in mobile and travel routers. WiFiRanger, Peplink, and Cradlepoint are all examples of mobile router brands that support WiFi-as-WAN.

Even the best router is limited by where you can place it. If it is stuck in a lower enclosed cabinet, or your RV or boat is built with metal walls, the router may actually perform worse than your laptop sitting on an exposed table near a window.

Routers and WiFi-as-WAN are discussed further in the **Routers** chapter later in this book.

Wi-Fi Range Extenders

More commonly available than WiFi-as-WAN routers are devices sold as Wi-Fi Range Extenders, though sometimes they are called Wi-Fi Boosters or Wi-Fi Repeaters.

These devices are designed to work as a relay to extend the range of an existing Wi-Fi network into dead-zones, such as the basement of a house or into a far bedroom.

To be effective you need to find some way to place the network extender midway between the Wi-Fi hotspot and the devices you want to connect.

In practice, this works well in a home with a room that has a weak signal, but is less practical for mobile users. These devices just don't have the range needed to reach distant Wi-Fi networks, and they assume you have control of the Wi-Fi network you're trying to extend.

USB Wi-Fi Network Adapters

When it comes to Wi-Fi range, nothing beats having a direct line of sight between the access point and your device with as few obstructions as possible.

If you have a window that faces towards the Wi-Fi hotspot you are trying to connect to, one of the most affordable ways to get a bit more range is to use a USB Wi-Fi network adapter with a more powerful Wi-Fi radio.

These are often marketed as "antennas" or "boosters", but if they are designed to connect to a single computer via USB they are actually fully self-contained external Wi-Fi networking cards. Similar to what is built into your computer, only with a much more powerful radio transmitter and a more capable antenna.

You can plug them into your laptop and set these small devices in a window facing towards the target hotspot. A few of these devices are even designed to be used outdoors, connected to your computer with a long USB extension cable to give you the ability to reach up to the roof.

The downside of USB adapters like these is that they will only get one single computer online, and they cannot help connect your non-USB devices to the distant Wi-Fi network. You may be able to buy a special companion router to work with your USB booster.

A lot of these devices have been slow to release driver updates to support the latest Mac OS and Windows releases too, so be careful to double check compatibility.

Roof Mounted Antennas & CPEs

Particularly with public Wi-Fi networks, a good signal is often very hard to find at ground level. If you want to connect over the greatest possible distance, nothing beats having long-range gear on your roof.

Having a roof-mounted Wi-Fi antenna paired with an indoor Wi-Fi router is especially handy if you are inside a metal signal-blocking tube such as a bus conversion, Airstream, or metal-hulled boat. You don't have to fidget with your gadgets trying to figure out which window has the best signal today.

Wi-Fi: Range Extending Gear

Unfortunately, Wi-Fi signals degrade rapidly when traveling over an antenna cable. Though it is possible to use an outdoor antenna connected to an indoor Wi-Fi radio via a coax cable, the amount of loss rarely makes this a good idea.

Instead of having the electronics indoors and the antenna outdoors, it is actually very common to mate the electronic brain of the Wi-Fi radio directly with the antenna in a single package rated for outdoor use.

This is referred as a CPE.

CPE stands for *customer premises equipment*, the common term used to describe the commercial-grade outdoor-rated Wi-Fi access points used by wireless service providers. A CPE combines roof-mounted height, better antennas, and a more powerful transmitter to deliver the maximum possible Wi-Fi range.

Odds are that any campground or marina with a professionally installed Wi-Fi network has CPE hardware mounted on poles or buildings around the park. You can use this same caliber of hardware on your end to ensure the best possible long range connection.

If you are comfortable configuring a device designed for network engineers and not average consumers, you can use a commercial roof-mounted CPE paired with any indoor router of your choosing.

For consumers, there are several companies that specialize in pre-packaged setups targeting the RV and marine market with custom firmware that takes care of the more complex setup. Many even offer customer support.

The upfront cost for going with a roof-mounted CPE setup may be steeper, but this is where the most substantial range and speed gains are to be had.

If long range Wi-Fi is central to your connectivity needs, it is often worth investing in doing it right.

For our guide to antenna installations:
www.MobileInternetInfo.com/antennainstall

Wi-Fi Is A Two Way Street

No matter how much money and time you invest in commercial-grade high quality CPE hardware mounted on the roof, the quality of the hotspot you are connecting to is usually outside of your control.

While some campgrounds and marinas will have also invested in long range outdoor CPE gear, many others have little more than a bargain-bin home wireless router from Walmart sitting behind a desk in the office.

That caliber of hardware is never going to be able to communicate over any substantial distance, no matter what you do on your end.

Sometimes you can take matters into your own hands, upgrading the hotspot you are communicating with to vastly increase your range.

Particularly when you are driveway surfing with friends and relatives, it can be very useful to set up a temporary high-power wireless access point in a window that is connected by ethernet to their home network.

For more on Wi-Fi and gear available, check our guide:
www.MobileInternetInfo.com/wifi

Satellite: Internet Everywhere

There is something magical and futuristic about being connected in the absolute middle of nowhere. Where only a satellite in space can keep you online.

As fun as it is to fantasize about connectivity everywhere, today's satellite options come with many tradeoffs to consider.

Compared to cellular service, satellite internet is often slower, higher latency, and more expensive. The gear to get connected can be bulky and requires setup at each stop.

If you plan to focus your travels on being way out in the boonies, the challenges of satellite internet might be worthwhile.

If you are hoping for a simple go-anywhere solution, satellite will likely frustrate you - at least for now.

Quick Glance

Pros

Signal (almost) everywhere

Cons

Latency

Data caps

Bulky equipment

Getting Internet from Space

It is important to make sure that you understand the basics of satellite.

Satellite TV vs Satellite Internet

Receiving a signal from space isn't particularly hard.

Sending a signal back to a satellite is where it gets tricky.

Satellite: Internet Everywhere

Satellite TV dishes are receive-only devices, and have no capability to transmit. Internet connectivity involves two-way communication requiring much more complicated gear.

Some people get confused because they see satellite TV providers like Dish Network and DirectTV (now owned by AT&T) advertising bundled packages that include internet service.

These bundled plans are intended for stationary satellite TV consumers to combine their TV, internet, and phone bills into one. The provider contracts out to local DSL or cable companies to provide internet service, usually relying on a hard-wired connection for internet, not satellite.

Even when the companies do offer satellite data plans for rural locations (like Dish's dishNET, which is provided by either ViaSat's Exede or HughesNet behind the scenes) they are strictly for fixed-location installs only.

In other words, they are not mobile friendly options.

> **NOTE:** Satellite internet systems are strictly for internet service, and are NOT compatible with satellite TV services. If you also want satellite TV to go along with satellite internet, you'll need a *second* dish!

Satellites Require Precise Aim

Communicating with a satellite over 20,000 miles away requires *extremely* precise aim.

For fixed installations with the dish mounted on a building, post, or even to a dock this can be handled just once, by a professional.

Mobile users still have options however.

1. **In-Motion Robotic** - It takes a large dome and some advanced and extremely expensive robotics to keep locked onto a satellite while in motion. Since a boat is never still this is usually the only option for marine - and is mega luxury yacht expensive.

2. **Auto-Aiming Robotic** - RVers who only need to connect while parked can install a roof-mounted satellite system that auto aims at the touch of a button. The dish retracts flat when not in use. The costs are more consumer friendly, but still pricey.

3. **Tripod Based** - The most affordable option is to learn to setup a surveyor's tripod to manually aim a dish at each stop. This can take up to 30 minutes, but has the advantage of allowing the dish to be located to avoid trees.

There are flat satellite antenna technologies in development that will allow for instantaneous electronic aiming. Until that military-grade technology becomes commercialized - big and bulky dishes are the current reality.

The Signal: Spot Beam vs. Broad Area

Older-generation communication satellites broadcast a signal that could reach an entire continent. This was great for mobile users. The satellite doesn't know or care whether you are in Boise or Boston, in the Black Rock desert or back-country Georgia.

But it is also horribly inefficient, with every user assigned to a particular satellite sharing the signal.

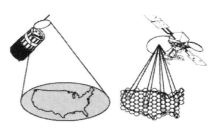

Newer satellite services use **spot-beam technology** to cover the nation with many small focused signals allowing for many more users to communicate at once.

You can think of each "spot" as being similar to a cell tower, only with the antenna located thousands of miles overhead.

Spot-beams allow for cheaper service and faster data speeds. Spot-beam also complicates things substantially for mobile users.

If you have spot-beam satellite service, if you travel very far from your "home" address, your satellite service will no longer work at all.

Why Not Just Change Spots?

It is technically possible for a satellite receiver to change spots, but very few consumer-focused satellite systems currently provide any support for automatically or even manually changing your assigned beam.

Satellite: Internet Everywhere

Relocating a dish to a new location often requires a certified installer aiming the dish while on the phone with the network operations center. You might only be allowed to relocate residential satellite service a couple times a year.

Because of the mobility-hostile nature of spot-beam service, it is quite difficult to take current *residential* satellite internet on the road.

Provisioning and training through commercial and business satellite resellers can get around residential consumer mobility restrictions. Certain HughesNet resellers can set RVers up to change spots on their own - which makes the newest Gen5 system with claimed speeds up to 25Mbps a compelling option.

Even Satellites Have Coverage Maps

The great advantage of satellite Internet service is that you can usually connect anywhere you have an unobstructed view of the southern sky.

But there are still coverage map issues when it comes to satellite.

Not every satellite serves every corner of the country, particularly if you want to travel to Alaska, Canada, down into Mexico, or cruise the Caribbean.

Spot beam services tend to be even more localized, with spots typically not aimed far beyond the national borders of the service's target market.

All geostationary satellites (in orbit over the equator) struggle to reach northern latitudes, making getting service in northern Canada and Alaska often a challenge.

It is smart to do some research in advance to make sure any satellite service you consider will support the place(s) you are most likely to go.

Latency & Satellite Communications

When you are using most satellite internet services, keep in mind that you are relaying everything via a satellite parked 22,236 miles above the equator.

That is a long way away, and the speed of light starts to impact everything in ways that feel completely foreign to those used to terrestrial connections.

Even with a fast broadband connection, there will always be a noticeable pause with every click as you wait for your request to make the trip from your dish, to the satellite in orbit, back down to the control station, then over the internet to the web server you are talking to, and then finally a full round trip back to you.

This delay is called latency, and it is especially noticeable if you try to have a video or audio conversation with someone over satellite.

Low Earth Orbit Satellite Constellations

Most current satellite internet options rely on carefully aiming at satellites in fixed locations in the sky parked in geostationary orbit over the equator 22,236+ miles away.

It usually takes a big, bulky, traditional-looking dish to communicate over those extreme distances.

Satellites in lower orbits can be easier to talk to, but they are always in motion through the visible sky and any given satellite will quickly pass over the horizon and out of sight.

But a constellation of multiple satellites working together can provide constant coverage from much lower altitudes, with orbits designed to ensure that at least one satellite is always in view at any time.

These are known as Low Earth Orbit (LEO) satellites.

LEO Tradeoffs

Because LEO satellites are in much lower orbits, it is possible to design a small receiver that does not need to be aimed, and which only needs a clear view of the sky to function. Which is a good thing, because constantly aiming at a moving target is a futile task.

LEO satellites allow for smaller receivers, and for much lower latency communications. They also support mobile users, and can offer true global service even at the Earth's poles.

But with current technology, LEO communication speeds tend to be extremely limited and data costs for more than just basic text messaging and tracking services tend to be prohibitive.

In other words, perfect for safety - awful for surfing.

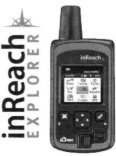

Satellite: Internet Everywhere

The key to using these expensive satellite services is extreme data-usage management, focusing on email and other text-centric communications rather than interactive graphical web surfing.

Next generation LEO satellites coming from Iridium, OneWeb, SpaceX, and others promise much faster speeds and much cheaper data, and potentially eventually even broadband speeds available anywhere on earth.

Realities: Today & Tomorrow

Current satellite options can provide an affordable and critical safety lifeline, and there are broadband speed mobile satellite options well within reach for those who truly want to go off the cellular coverage map.

But for most nomads, the advantages of satellite will not outweigh the challenges quite yet. Both cellular and Wi-Fi are typically just much easier to deal with.

And even once the next generation satellite systems are fully deployed, they are in most cases being designed to be complimentary systems to terrestrial cellular, not replacements.

In an urban or suburban area satellite will be hard pressed to compete with fully built out LTE on the ground.

On the other hand, in many remote areas it will never make sense to fully build a network of cell towers. No matter how much cellular companies expand, there will always be gaps in coverage.

The ideal connectivity future involves a mix of easy to deploy satellite and cellular, with service roaming seamlessly to the best connection possible wherever you happen to be.

It will take a while for all the essential pieces to fall into place, and for the necessary partnerships and technologies to emerge.

While we wait – we can look to the skies and dream!

Our Guide to current and future Mobile Satellite Internet options:
www.MobileInternetInfo.com/satellite

Thinking Outside the Box

Cellular, Wi-Fi, and satellite are not the only ways to get online while enjoying life on the go. If you are willing to get a little creative or experimental, there are a few other less obvious alternatives too.

Particularly if you are willing to get a little flexible with what "mobile" and "internet" means, you may have more options than you ever imagined.

Cable / DSL / WISP Installation

Being mobile doesn't necessarily mean having to use only mobile internet.

If you're planning to be in one spot for a while, sometimes hooking up directly to traditional wired cable or DSL is a possibility.

RV parks and marinas that cater to long-term residents sometimes already have cable available, and all it takes is contacting the local provider to get a modem and subscribe.

They may not advertise this as an amenity, so it's worth asking. Start by looking for places that offer cable TV or seasonal rates.

Depending on the provider, the costs to get started are often very reasonable.

There are not often long-term contracts so you can cancel after just a few weeks without penalties.

You can usually rent the equipment for a few dollars a month, instead of buying. If you find yourself signing up for cable internet often, many providers utilize the same modem standard, so it may be worthwhile to buy a cable modem and keep it onboard for quicker activation.

The advantage of going with cable or DSL is gaining access to fast and essentially unlimited internet. After rationing out bandwidth while mobile for months on end, spending a few weeks drinking from the firehose can feel incredibly decadent!

Temporarily embracing a wired lifestyle can sometimes be very worth it for us bandwidth junkies and perfect for hyper-focusing on a work project before heading back out for new adventures.

WISP Access (Wireless Internet Service Providers)

A WISP is a wireless internet service provider. These companies have sprung up in many communities to fill the demand for faster-than-dial-up home internet service. The WISP providers set up transmitters on local high points and then install compatible broadband receivers on the roofs of local customers.

Since the WISP doesn't need to run new wires, if you have a view to the right mountain or hillside, you might be able to get a local WISP to offer you fast unlimited service, even if you are in a remote boondocking spot miles from the nearest cable run or phone line. It all depends on line of sight.

To find out if a WISP might be an option in your area, check local advertising periodicals, signs in grocery stores or laundromats, the Yellow Pages, Google, or talk to local computer repair professionals for leads.

Because a professional installation is usually required, WISP service is not usually appropriate for short-term stays.

Co-Working Spaces

Co-working spaces are office setups for independent workers to base themselves out of. They offer amenities like a desk, connectivity, and meeting rooms. Sometimes they can be far more productive spaces than trying to get work done from a cafe, as everyone around you is also working.

Thinking Outside the Box

Co-working spaces are now located all over the country, particularly in larger cities. Some of them offer short term rental options including hour and day passes, or even monthly space. These can be an ideal option for a mobile worker to setup a work base camp for a bit.

Start with resources including CoWorking.org and ShareDesk.net to locate available co-working spaces.

Amateur Ham Radio: Email & Internet

If you're a licensed ham radio operator with the right equipment, you can get access to Winlink.org.

This service allows you to send noncommercial emails over the amateur radio frequencies, which can be useful if you're out in the middle of nowhere with no other options.

The important thing to remember about amateur radio is that it is strictly for amateur use. It is against the law to use the amateur radio bands for any commercial purpose (including checking work email) or to transmit any encrypted data, so doing online banking is out.

It also means that, for the most part, you need to be comfortable figuring things out on your own.

If tinkering with radios turns you on, this can be a great free way to get online. Even with the inherent limitations, as a ham you will be able to manage some very basic data connectivity on the go, wherever you are.

Broadband-Hamnet (hsmm-mesh.org) is an experimental, high-speed, wireless broadband network that uses Wi-Fi networking gear with custom firmware to build a self-configuring mesh network. If you are in an area with other users, you can be part of creating a widespread network that stretches much further than a single user alone could ever communicate.

If you want to get involved with amateur radio, the place to start is the ARRL - the National Association for Amateur Radio.

Concierge Services and Personal Assistants

Sometimes it's best to let someone else do your tasks for you.

Think about how long it might take you to research a topic or make a reservation with a web connection that bounces up and down.

What if instead you had someone with a rock-solid internet connection willing to do your bidding on demand, and who will get back to you later with the results or needed information?

Tap into Online Community

Rather than fighting with a slow connection for hours, try instead to get online just enough to post a request for advice and information to your Facebook friends, your blog, or to to a relevant forum you participate in.

This is a great way to get routing recommendations or general technical advice. And, of course, return the favor for others when you have a good internet connection and someone needs a little information.

Friends / Family

If you have specific requests and tasks that need to get handled and your lack of good connectivity is making you pull your hair out, don't be afraid to get a message out to a friend or family member asking for help, particularly if you are dealing with an urgent or emergency situation.

Credit Card Concierge

There is a chance that you might already have a personal assistant on call and not even know it!

Some credit cards (such as Visa's Signature cards and many AMEX cards) offer a concierge perk, and it can actually be surprisingly useful.

Online Virtual Assistant Services

If you have tasks that are more complicated than a free concierge can handle or you want a concierge who "knows you," there are many online personal assistant and concierge services that you can subscribe to that can handle both on-demand and recurring tasks.

Some nomads have even found it worthwhile to actually have someone on staff dedicated just to their needs.

If you are running a business and need to keep up appearances of always being available, having someone who answers the phone, triages replies to emails, and handles making reservations for you can buy a lot of flexibility and freedom.

Routers: Bringing It All Together

Routers serve as the central conductor on any network, acting as a gateway between the Local Area Network (LAN - your devices) and the Wide Area Network (WAN - 'the internet').

If you want to have more than one device taking advantage of a single upstream internet connection, or you want to connect your local devices together to share files or functionality, you need a router.

Typical residential routers often connect to a cable or DSL modem for the WAN uplink, and create a local Wi-Fi and wired ethernet network for all the local devices in a home to connect to and share this upstream connection.

But on the road or water, cable modems and DSL lines are rarely found.

Instead, upstream connection options are frequently changing between cellular modems and public Wi-Fi hotspots, and it takes a special kind of router to be able to interface with and switch between these connection types.

The most basic cellular router is a Mobile Hotspot or MiFi or Jetpack, as discussed in the **Cellular Data: Gear** chapter. A smartphone creating a Personal Hotspot is doing the same, acting as a basic router and sharing its cellular connection with other nearby devices.

If you need more capability or

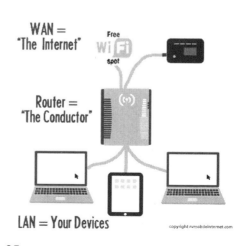

flexibility than a mobile hotspot can provide, more advanced router options are out there, allowing you to connect to both cellular and remote Wi-Fi networks as upstream WAN sources, and these routers make it easy to change between upstream connections too.

You can even have WAN inputs like satellite, cable, or DSL added as options should you have access to them.

Some routers even let you connect to and combine multiple WAN networks (i.e. two cellular networks, or a cellular connection and public Wi-Fi) at the same time for instant failover, or for increased speed or reliability.

The core features that set mobile-friendly routers apart from home routers is support for at least some of the following features:

Embedded Cellular Modem: A router with a cellular modem integrated into it can use a SIM card and an active data plan to connect directly to a compatible cellular network with no additional hardware needed.

USB Cellular Modem Support: A router with a USB port that can control a compatible USB cellular modem or tether to a cellular mobile hotspot allows for a cellular connection to be shared.

WiFi-as-WAN: A router with this feature can connect to an external Wi-Fi network upstream (such as a campground or marina Wi-Fi hotspot), and at the same time create a private Wi-Fi network downstream for your local devices to share that connection. This allows all of your personal devices to always connect to the router, and lets the router worry about what is the connection that best works at your current location.

Flexible Power Inputs: Many mobile users want to be able to optimize for 12v power to run off their house battery systems. A hallmark of mobile routers is the ability to run off 12v directly.

Do You Really Need a Router?

The most basic function of a router is taking an upstream network connection and sharing it with multiple downstream devices over either wireless Wi-Fi or wired ethernet.

If you never intend to share a single connection across multiple devices, you do not need a router. A simple USB modem plugged directly into a laptop or a hotspotting off a phone or tablet can save you headaches.

Use Cases

And if you only have a few devices that you want to share a connection with, you can probably just use the routing capabilities built into the personal hotspot feature on your smartphone or a dedicated mobile hotspot.

If, on the other hand, you have an entire collection of devices that you'd like to get online, potentially via multiple upstream connections, as well as enabling your devices to talk to each other, you almost certainly would benefit from having a router sitting at the heart of your network.

In some mobile homes, everything from the lightbulbs to the bathroom scale are Wi-Fi enabled, and a router is an absolute necessity!

Router Selection Tips

Some nomads use traditional home wireless routers, such as the Apple AirPort or any of the vast range of routers sold by LinkSys, D-Link, Netgear, and many others.

These home routers, however, do not have any built-in support for cellular data connections, making them substantially limited for mobile use.

Some general consumer routers and Wi-Fi range extenders support Wi-Fi repeating and a whole raft of other features. But they are all designed for extending Wi-Fi networks where you control both ends of the connection, a situation that is hardly ever true for most travelers.

You will save yourself some headaches by going with something designed with the mobile user in mind.

Cellular & WiFi-as-WAN Routers

Though it ends up being a more complex setup than a MiFi, often there are advantages to using a specialized router that can bring together multiple upstream connection options.

The key feature to look for is direct support for controlling internal or tethered cellular modems, and for a feature known as WiFi-as-WAN. As mentioned earlier, this is the capability to simultaneously create a local private Wi-Fi network while also connecting upstream to a public Wi-Fi network.

Use Cases

A mobile router may also be able to juggle other upstream data sources, such as a cable modem or satellite connection, and you should look for failover features that define how these connections are prioritized and automatically switched between.

A common configuration is setting up your router to first connect via WiFi-as-WAN to an open Wi-Fi network, and then automatically failing over to cellular. This way the free and unlimited connection becomes the priority.

The nicest thing about using a router like this in the heart of your network is that it keeps things simple on all your client devices. You just need to configure all your devices to point to your router's hotspot, and when you change locations you do not need to reconfigure any of your devices or reenter any new passwords.

Wired vs. Wireless Speeds

Plugging your fancy laptop into a wired network jack may seem like a step backwards into the primitive pre-wireless age. But, if you have multiple computers talking to each other (and not just upstream to the internet) it can actually end up making a lot of sense.

Consider, even if you have a 5GHz 802.11n wireless network, the maximum theoretical speed is usually at best 300Mbps, and in less-than-ideal conditions, the speeds will be much, much lower.

Even the latest 802.11ac Wi-Fi will struggle to keep up with a fast wired network, especially if there is a lot of local wireless network traffic.

Meanwhile, a wired gigabit ethernet network can run at a full 1,000Mbps speed in both upstream and downstream directions simultaneously.

If you keep your backups and media stored on a network attached storage (NAS) drive, having a wired network means that these drives are reachable from the other computers nearly as fast as a local hard drive would be.

If you do build a wired ethernet network, seek out hardware that supports gigabit speeds. "Fast ethernet" can barely outrun 802.11n Wi-Fi.

If you need more wired ethernet ports, you can just add a cheap gigabit ethernet switch to the network without impacting performance at all.

Our deeper Guide to Mobile Routers and Options:
www.MobileInternetInfo.com/routers

Use Cases

What you need or want to do online will have a huge impact on the mobile internet setup that best keeps you connected.

We went over some of the general considerations in the **Assessing Your Needs** chapter at the beginning of the book, but now that you've had a chance to read about some of the options, we'll go over some specific considerations for some common use cases

Every one of these topics could be a book of their own, so definitely click through to the linked guides which contain more in-depth details on each.

Working Remotely

If you'll be working remotely online, then mobile internet isn't just a "nice to have." It can be absolutely essential. Your livelihood may be dependent upon your ability to get connected to get your work done.

Assessment

There is a lot of variability in the requirements for online work.

Do you need to be connected during specific work hours or have frequent scheduled conference calls?

Or do you have a flexible work life where your deliverables matter more than the specific hours you happen to be online?

Use Cases

How data intensive are your work tasks? Someone who has frequent video conference sessions will have much higher bandwidth needs than someone who just needs to process e-mails and text based documents.

And, those dealing with client sensitive data, such as medical or financial documents, might have extra privacy and security considerations.

Work requirements can impact the speeds you need to be able to maintain, the amount of data you need to be able to tap into, the security measures you need, and the reliability of your setup.

Redundancy

Redundancy is having multiple options on board for internet access, and it is a key to building a reliable mobile internet solution.

- Have data options with two or more cellular carriers on board (we often see folks with 3+).

- Have a range of cellular signal enhancing gear available, and know when it makes sense to invest time in optimizing your connection.

- Keep options for gaining access to public/private Wi-Fi hotspots.

- As a backup plan, be open to seeking out a cafe or coffee shop for a little free Wi-Fi with your lunch.

Planning Around Connectivity

The more schedule oriented your work life is, the more important it will be to scout out connectivity while planning your future stops.

As we discussed in the **Going Wireless** chapter, use resources like review sites and coverage map apps to know your connectivity in advance.

Save yourself a lot of stress and arrive with plenty of time before a big work day or deliverable. This gives you time to test out all your redundant connectivity options so that you have your primary and backup plans ready.

There is nothing more frustrating than pulling into your new location after a long repositioning day right before a scheduled conference call, only to find a weak signal you weren't anticipating that derails everything!

For our guide to Mobile Internet for Working Remotely:
www.MobileInternetInfo.com/work

Families & Homeschooling

Thank you to Andrew McNabb for providing this section and accompanying guide.

There are some unique mobile internet challenges that can impact families on the road. Everything from keeping the kids entertained, to controlling and managing internet access, to ensuring you have the internet options you need for critical family tasks like homeschooling and working online.

Or even giving the adults a break while the kids watch Netflix.

Many kids are growing up with technology and are completely comfortable with the internet. After all, they don't have any experience living in a world without it. As a result, kids can be voracious consumers of internet content if left unchecked, and won't initially understand that mobile internet on the road or water can come with limits.

Homeschooling

Nomadic families with school-age children use alternatives to public schools and commonly this means some type of homeschooling program.

There are many programs to choose from, and internet requirements vary widely. Keep these things in mind:

- **School IT requirements:** Some schools utilize or require a lot of video streaming or video conferencing, which will use a lot of bandwidth. Be sure to research the program to see how much bandwidth use is typical.

- **Schedule requirements:** An online program with a fixed schedule will require stricter travel planning to ensure internet access is available when needed. A flexible schedule provides more adaptability for varying internet access.

Managing Family Access

Sometimes you may only have an internet trickle and every member of your family will want it all!

Parents may also want to control or manage access to specific content, or block access at specific times of day, or block some devices from the internet while allowing others access.

A family will likely have many devices requiring internet, and may have multiple mobile internet connection sources.

- **Centralize control with a mobile router**: Management of internet devices and internet sources can be much simpler through the use of a mobile router. See our chapter on **Routers** for more info.

- **Manage access like a library:** This is essentially a parenting strategy that's been used for eons. Control access to the devices themselves and check them in and out as needed.

- **Self-Management:** This strategy gives an individual their own internet source and they manage it themselves - dealing with the fallout from running out of data or being out of connection as they may.

For our Guide to Families on the Road:
www.MobileInternetInfo.com/families

Video Streaming: TV & Movies

While nomads do tend to enjoy getting out and exploring our new locales, when travel is a lifestyle and not just a vacation, it's all about finding balance.

Many of the traditional ways of staying entertained just don't work like you might be used to on the road.

The cable is cut, satellite TV is hit-or-miss through trees or other obstructions, over the air channels are completely unfamiliar, and online streaming can leave you fighting with flakey connections and living in fear of hitting your cellular data caps.

Nothing consumes data quicker or is more demanding than streaming video!

You can easily blow through your data bucket for the month quicker than you might think possible.

But, it is possible to get your entertainment fix if you put some thought into your strategy in advance.

Using Mobile Internet for Streaming

You're not going to often find free public Wi-Fi (such as that at campgrounds or marinas, or even at stores and cafes) that is capable of reliably streaming video. In fact, many of these provided resources actually restrict video streaming on their networks because it is so bandwidth intensive and can prevent others from doing basic web surfing.

So if video streaming will be important to you, be prepared to bring your own data plans - usually over cellular.

All of the carriers now integrate in the option for video streaming into their plans. Generally, these options for "unlimited video streaming" are intended for watching video on the small screen of your smartphone or tablet.

This may be fine for some, but you might want to watch your favorite shows on the big screen.

For that, you have two options:

- Get a cellular plan with unlimited mobile hotspot use so you can use streaming devices like smart TVs, Apple TV, Chomecast, etc.

- Utilize your plan's on device unlimited data with screen casting or HDMI out options, which use the device's on-device unlimited data and just mirrors the image to the big screen.

Alternatives to Streaming

If streaming live over the internet will not work with your travel plans or mobile internet setup, then pre-think options for viewing content.

This could include:

- **Download content offline.** Several video services allow customers to download content to their mobile devices for viewing later.

- **Rent DVDs and Blu-rays.** If you're staying places for more than a few days, Netflix's disc rental service can work quite well on the road. And RedBox rental kiosks are often located in convenient locations for renting a recent movie for a night. You can also often even borrow discs from libraries.

- **Rip content to a hard drive.** Keep a legally obtained DVD collection in storage, and keep ripped backup copies stored on a portable hard drive.

- **Buy TV series on disc.** Purchase your favorite series on disc. Or, you can often find series discs at low cost used. You can also borrow from libraries or trade with other nomads.

- **Tuner and DVR setup.** There are TV tuners that can attach to an external TV antenna and can pick up local stations on your computer and then save the content to a hard drive to watch later.

- **Over the Air antennas.** If you like watching local TV stations, this old fashioned way of watching shows and news still works on the road with an antenna.

- **Satellite TV.** Install and subscribe to satellite TV services.

The best options for streaming on the go are always changing, and we track them regularly online.

For our guide to TV, Movies & Entertainment on the Road:
www.MobileInternetInfo.com/tv

Online Gaming

Thank you to Andrew McNabb for providing this section and accompanying guide.

Almost all games now come with an online play mode, and many games today require internet connectivity even for single-player gameplay as well as for patching and authentication.

Gaming connections generally require a consistently higher quality connection than many other internet uses, which can be a challenge.

Use Cases

Although there will be times when some games can't be played, or can't be played enjoyably, in many cases mobile online gaming is possible.

Some of the challenges of using mobile internet for gaming include:

- The variable latencies associated with mobile internet may be a problem for some game types.

- Carrier grade NAT on cellular connections may require use of a VPN for some gaming.

- Games can require large updates that you have little to no control over in terms of scheduling the download.

Here are some tips and tricks for dealing with some of the challenges of gaming while using mobile internet:

- Build redundancy into your mobile internet set-up. Players with data caps should consider alternatives for downloads and updates like borrowing a friend's high-speed home internet connection or utilizing public Wi-Fi.

- Exercise extreme caution when connecting a gaming platform directly to a data-capped hotspot device or tethered smartphone.

- If latency is an issue, play a game that doesn't require low latency. Online board games, turn-based strategy games, 4X, online collectible/trading card activities, or LAN based games are all playable even with latency issues.

For our Guide to Gaming over Mobile Internet:
www.MobileInternetInfo.com/gaming

International Travel

US based nomads don't just stick to the US for their adventures.

Whether via RV, van, boat, or ditching your typical mode of nomadism and utilizing sticks and bricks accommodations, many of our mobile brethren will cross into international territory at some point - and will want to keep connected.

The hurdle with international internet is not that other countries don't have plentiful options. It's getting connected to them as nonresidents who are just passing through the country on a short-term basis.

Here's a brief summary of the options for keeping connected when you travel abroad, with our specific guides listed at the end.

Wi-Fi

Public Wi-Fi is plentiful abroad, just like most other basic necessities of life.

You will often be able to connect at campgrounds, marinas, coffee shops, cafes, libraries, hotels, airports, and more. You will be surprised just how plentiful basic Wi-Fi can be, even in otherwise primitive countries.

When traveling outside the USA, Wi-Fi is likely going to be your cheapest and easiest connectivity solution, especially if you're only going to be in an area for a brief time when it may not be worthwhile tracking down other options.

International Cellular Roaming

Most US cellular plans include some international support and options. Know your carrier's limitations and costs, as well as any options you might need to turn on in advance of crossing the border. Generally, there are different options available for North America roaming (Canada / Mexico) versus the rest of the world.

Roaming with your home carrier may be ideal if you're only planning a short trip, will primarily be relying on Wi-Fi, or won't be needing much data to get by.

For longer trips or heavier needs, it often makes sense to look into other options.

Global Cloud SIMs

There are companies that provide global hotspots for traveling far and wide. This option is of particular interest if you will be hyper-mobile, bouncing from country to country over a short period of time (such as on a commercial cruise vacation).

Global SIMs provide easy global support, with no need to purchase SIM cards in each country.

Typically, when you're abroad, you will pay a single flat fee for a set number of hours of access, which will provide you with a set amount of data (often unlimited) for that period.

Going Native: Getting a Local SIM

If you're planning more extended time in a country, it may be worthwhile to seek out options with the local carriers to get a local prepaid or no-contract cellular plan.

Most of the rest of the world has standardized on GSM technology for their 2G cellular networks, and if you have an unlocked phone, tablet, or hotspot that is global GSM compatible (almost all are) you will be able to pick up a SIM card in many countries. By putting the local SIM in your device, you can make calls and surf the internet from your own tech at local rates. Make sure you have unlocked your device with your carrier however, or a SIM from another carrier will not work.

And if you want fast LTE speeds, make sure that your device supports the LTE bands in use in the country you are visiting.

For our guides to International Mobile Internet:

General: www.MobileInternetInfo.com/international
Canada: www.MobileInternetInfo.com/canada
Mexico: www.MobileInternetInfo.com/mexico

Staying Safe & Secure

Though this is an important topic for everyone, mobile users face a few unique challenges in staying safe and secure online.

One particular area of concern revolves around staying safe while using public Wi-Fi hotspots. These networks are shared resources, and if they are not protected with a network password anyone nearby can listen in on any non-encrypted network traffic.

And even if there is a password, do you really trust all your marina or campground neighbors?

Use Cases

Also, if your computer is configured to share files, music, or photos on the local network, by connecting to a shared Wi-Fi network you might be revealing your musical tastes or other even more embarrassing details to anyone who cares to look.

Cellular networks are inherently much more secure than public Wi-Fi since there is no one else sharing your network, but many people are still concerned about their surfing habits being monitored by their carrier and used for advertising and other marketing purposes.

Using a VPN (Virtual Private Network) adds a strong layer of security that encrypts all of the traffic between your computer and the VPN server. But, a VPN adds costs, complexity, slowed speeds and latency to every connection.

Another concern for mobile users revolves around keeping backups safe.

Hard drives will inevitably fail. It is not a question of if, but when. And when mobile they can very literally crash – headlong into another vehicle.

Having good backups is thus doubly essential.

Cloud backups typically use too much data to be reliable for heavy users, and backing up to a spare hard drive stored in the same vehicle does nothing to protect against fire, sinking, or vehicle theft.

Everyone will need to strike their own personal balance between security, cost, complexity, and convenience.

For our guide to Mobile Internet Security:
www.MobileInternetInfo.com/security

For our guide to Back-Ups over Mobile Internet:
www.MobileInternetInfo.com/backups

Wrapping Up: Top 10 Tips

There is a LOT of information in this handbook. To help wrap things up, here are some essential words of wisdom distilled down to their core.

Tip #1: Redundancy! Redundancy!

The more possible onramps to the internet at your disposal, the more likely you are to find one that works where you are at. Embracing a diversity of connection types is the best possible way that you can maximize your chances of getting at least a somewhat workable connection.

Tip #2: Soak up any (free) Wi-Fi you find!

Sometimes the fastest, cheapest, and easiest way to get online is to use public Wi-Fi networks. Use Wi-Fi when you find it! But, don't count on it – usably fast and free Wi-Fi is an unfortunately rare combination.

Tip #3: Understand roaming & coverage issues!

Because they want their networks to seem as large as possible, carriers go out of their way to hide that you may actually be roaming and subject to running into usage limits. Stay on guard!

Tip #4: Be aware near borders!

Beware of international borders! Older cellular plans can charge an arm and a leg for international roaming (including onto the onboard cell networks offered on cruises). If you are going to be anywhere close to an international border, make sure to turn off data roaming on all of your devices unless you know your plan is setup to roam for free, or with an affordable Day Pass.

Tip #5: Know your limitations!

Most fixed-location internet connections are unmetered, but mobile data can have limitations including speeds being throttled after a certain amount of usage, or expensive overage charges on older legacy plans. Save your big downloads for the days you have truly unlimited access.

Tip #6: Re-assess your setup annually!

Mobile technology changes often. The carriers regularly come out with more attractive plans or promotions, and it's important to look over your plan at least annually to see if you're getting the best pricing and options. Cellular devices (phones & hotspots) in particular should be upgraded at least every 2 years to make sure they can connect to the latest frequencies.

Tip #7: Redundancy!

Need we say it again? Yes, redundancy is that important! Having access to multiple cellular carriers can be particularly valuable, and is usually worth investing in.

Tip #8: Focus on Performance, Not Bars!

A signal booster and/or an extendable antenna mast can work wonders to help you get online from afar. These systems aren't magical, but occasionally they can make the difference between having a barely detectable signal and a barely usable one.

But don't focus on just how many bars of signal you see. Test your connection to see the actual performance increase!

Tip #9: Stay safe out there!

The internet is a scary place, and public networks can be especially so. To keep yourself safe, never ever use the same password in multiple locations. To take your security and privacy even further, subscribe to a VPN service.

Tip #10: Manage Your Expectations!

Planning in advance on having good net days and bad net days (and even no net days) is perhaps the ultimate key to avoiding frustration!

Go Further . . .

Congratulations. You've gotten through this book, which is a basic start to understanding the challenges of mobile internet and the options out there.

But don't let your learning stop here!

There is so much more to know about mobile internet to keep successfully connected, and we could easily turn this into a 500+ page book that would completely overwhelm most.

Not to mention, the details change so often we'd have to rewrite it monthly.

Instead, we have comprehensive living guides online that are awaiting you when you're ready to dive in deeper at your own pace. These guides include details on specific current products, plans and options. We're constantly keeping these guides updated as technology shifts.

Here's our bookshelf of popular guides to start with:

Our Featured Guides:
www.MobileInternetInfo.com/featured

We are always expanding our library with new guides, videos and resources as we track the industry, test gear and interact with our fellow nomads on this topic. We try to offer a lot of it for free, but some is exclusive for our members.

Our premium membership also includes an organized classroom to walk you through our guides and video content to continue your education - as well as regularly scheduled interactive webinars.

Get a sneak peak at Mobile Internet University:
www.MobileInternetInfo.com/classroom

It is with huge gratitude to our premium members for funding our resource center, as books sales alone don't cover the efforts to keep these guides constantly updated and expanded.

Mobile Internet Resource Center

In the time it took you to read this book, we promise you an industry change has occurred that has already made something in here out-dated.

We're constantly staying on top of this topic at MobileInternetInfo.com, the website meant to complement this book. We analyze tech news for how it relates to mobile internet and nomads, as well as provide more in-depth guides, product reviews, videos and more.

Mobile Internet
RESOURCE CENTER
mobileinternetinfo.com

Check for free periodic wrap-up articles that update this book:
www.MobileInternetInfo.com/changes

Other places you can keep updated:

- Join our free monthly newsletters:
  www.MobileInternetInfo.com/subscribe

- Join our free public Facebook discussion group:
  www.facebook.com/groups/rvinternet

- Follow our free Facebook News Page:
  www.facebook.com/MobileInternetInfo

- Add our news feed to an RSS reader:
  www.mobileinternetinfo.com/feed/

- Subscribe to our YouTube Channel:
  www.youtube.com/MobileInternetResourceCenter

Mobile Internet Aficionados: Our Premium Membership

If this book is the textbook..

...the MIA is the classroom.

Our premium membership is designed for those who depend on mobile internet to enable their roaming lifestyle.

Our goal is to save you time, money and frustration.

We keep on top of the industry -
so you can focus on what drives you!

Our premium MIA members get:

- Q&A forums – where you can ask us questions and get answers.

- Mobile Internet University - our self-paced classroom to walk you through our in-depth content.

- Alert newsletters – more timely and in-depth than our free monthly newsletter, bringing you up to date on any industry developments that might impact you, as well as insider scoops.

- Exclusive in-depth content released to members first – such as reviews, equipment testing results, and advanced guides.

- Interactive topical and Q&A webinars.

- Special vendor discounts that could easily save you more than the cost of membership.

- Gratitude from the nomadic community for helping us provide the free & non-sponsored public resources we're able to offer.

- *Free Updates to this book* in PDF format.

If this stuff is vital to your mobile lifestyle, we invite you to join us at:

www.MobileInternetInfo.com/membership

Save $5 on Membership

Our Mobile Internet Aficionados premium membership includes a free copy of this book in PDF eBook format, and any updates that might be issued during their membership term.

To thank you for already having purchased this book separately, we offer you $5 off the price of a new membership to the MIA.

When joining at www.MobileInternetInfo.com/membership, use this discount code to save $5 at checkout:

> ## Join the MIA
>
> Save $5 off a new membership.
>
> Use Coupon Code:
> **TMIH5th**

In Thanks

We can't do this all alone! We'd like to take a moment and extend our sincerest gratitude towards:

Our team - Liz, Andy and Tom have been our rocks this past year, helping us keep an eye on the industry, writing stories and guides, editing our writing (extra thanks to Liz for editing this book!) and assisting our nomadic communities in our Internet for RVers & Cruisers Facebook group. And of course, providing comic relief on the extra busy days.

To our cats - The masters of us all, thank you for giving us some time off of our cat worshipping duties to get our work done.

Our families - Thank you for your patience as we worked on this book over our holidays and shared vacations together (much of this book was written on a cruise ship). And, of course, for being our biggest cheerleaders in the pursuit of our dreams!

Hurricane Irma - Thank you for sparing our homes, both of which were parked in your direct path. Without your visit, however, this book would have been published back in September and would already be outdated by the beginning of 2018. Quite honestly, the extra time gave us an opportunity to really think through this total re-write, so thank you for forcing us to do it right.

Our members - We've said it many times throughout this book, but we'll say it one more time. Without the funding of our Mobile Internet Aficionados, there is no way we could afford to make this topic our professional focus and keep updating this book. The MIA has grown and grown since 2014 and has become a vibrant community of thousands of nomads who share their experiences and assist each other. There are only so many hours in the day and so much space in our mobile dwellings to install and test gear; the collective wisdom and experience of our community is the greatest asset we learn from.

Made in the USA
Lexington, KY
14 June 2018